The Universe, But This Time You Understand It.

Cell phones on silent,
the show is about to begin

AASHNI JOSHI

notionpress.com

INDIA · SINGAPORE · MALAYSIA

Copyright © Aashni Joshi 2022
All Rights Reserved.

ISBN
Paperback: 979-8-88733-392-2
Hardcase: 979-8-88805-474-1

To my family who supported me at every stage,
To my teachers who nurtured my curiosity and
To my friends who were always there for me.

CONTENTS

EPISODE 1:
THE BIG BANG

Just like story books, this book also starts off our journey from the very beginning. How it all began. How it all came down to a snap. The birth of the observable universe as we know it. The Big Bang.

The Big Bang Theory is the most widely accepted theory on how the universe began. Simply defined, it states that the universe as we know it began with an unimaginably hot and dense single point that inflated and stretched — at first at unimaginable rates, then at a more measured rate — over the next 13.8 billion years to become the still-expanding cosmos we see today.

Let's take a time-machine and travel 13.7 billion years into the past. Everything in the universe was condensed into an infinitesimally small singularity, a point of infinite density and heat.

https://earthhow.com/history-of-the-universe

Suddenly, our universe began to expand at a rate greater than the speed of light, expanding outwards faster than the speed of light. This was a brief moment of cosmic inflation, lasting only a fraction of a second (approximately 10-32).

The universe was extremely hot at the moment, churning with electrons, quarks, and other particles. All the universe's ingredients were there but it was still too hot and dense for subatomic particles to form.

Following that, the universe cooled enough for quarks to fuse together and produce protons and neutrons at 10-32 seconds. The protons and neutrons then underwent a dramatic transformation approximately a second after birth. They merged to form nuclei. **This is known as the nucleosynthesis era**. Some of these nuclei also joined to make helium, however in far lesser amounts (just a few percent)

However, nucleosynthesis stopped after roughly 20 minutes, and no further nuclei could be formed. Because of the massive heat and radiation still flooding the universe, electrons couldn't stay in orbit around any atomic nucleus at this stage. They were knocked apart again by intense radiation shortly after any neutral atoms formed (neutral atoms just have the same number of protons and electrons and hence bear no overall charge).

After about 380,000 years, the universe had expanded and cooled sufficiently to allow electrons to remain in orbit around atomic nuclei.

About, 380,000 years after the Big Bang, this allowed light to eventually shine through.

https://www.forbes.com/sites/jamiecartereurope/2020/05/14/is-it-time-to-dethrone-the-big-bang-theory-heres-how-you-overthrow-it/?sh=6d24aafe20d9

This is when recombination took place, allowing neutral hydrogen (and helium) to "recombine with" (hang on to) electrons without being easily lost to stray radiation.

Because it couldn't hold visible light, this early "soup" would have been difficult to perceive. According to NASA, "the unbound electrons would have caused light (photons) to scatter in the same manner that sunlight scatters from water droplets in clouds."

However, over time, these free electrons interacted with nuclei, forming neutral atoms, which have equal positive and negative electric charges.

At last, there was light! Alright, but what about the stuff we skipped over? So, what happened right at the start? We have no idea what occurred here.

Our tools have abandoned us at this point. Natural principles cease to make sense, and time itself begins to wobble.

We need a theory that integrates **Einstein's Relativity** with **Quantum Mechanics** to understand what happened here, which countless scientists are now working **on.**

However, we are left with so many unsolved questions.

+ Were there other universes that existed before ours?

+ Is this the universe's first and sole instance?

+ What caused the Big Bang, or did it happen naturally, according to laws we don't yet understand?

+ We don't know, and we may never find out.

+ But we do know that the universe as we know it began here, by giving birth to stars, galaxies, planets, the Earth, and you.

Speaking of stars, let's get right into it.

EPISODE 2: STARS

We've often heard the nursery rhyme — "Twinkle twinkle little star, how I wonder what you are.?" But guess what? I'll tell you what they are. Stars are a big ball of gas.

Like our sun, there are millions of stars in the universe.

A major part of a star's life is pretty boring. They live for billions of years without a lot of major activity. Let's take a look at how these gigantic spheres come to life.

A star is born within cold, dense molecular clouds in space called nebulae. These nebulae contain mainly hydrogen and helium. Certain events such as a huge supernova explosion can cause this cloud to collapse under its own gravity. This leads to hydrogen and helium atoms clumping together at the centre of the cloud. This hot core eventually forms what's called a **PROTOSTAR.**

You may wonder: if this force of gravity has collapsed such enormous clouds, then how does a star sustain itself? The answer is **NUCLEAR FUSION**.

The star at its hot core is continuously fusing hydrogen into helium to release enough outward energy to counteract this strong force of gravity.

Just like Rajesh Koothrapalli from The Big Bang Theory said, "And with apologies to Lady Gaga, th how a star is born."

When temperature and pressure further increase, over millions of years, we finally get our main sequence star.

Like Sherlock Holmes's magnifying glass, the **NASA/ESA Hubble Space Telescope** can peer into an astronomical mystery in search of clues. The

This image covers a portion of a large galaxy census called the Great Observatories Origins Deep Survey (GOODS). *Credits: NASA, ESA, the GOODS Team, and M. Giavalisco (University of Massachusetts, Amherst)*

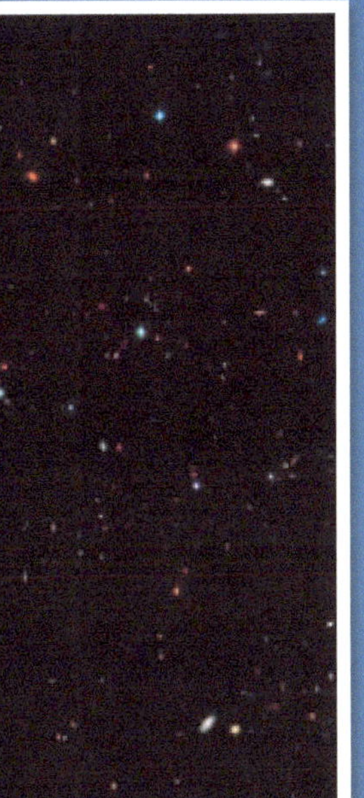

enigma in question concerns the globular cluster Ruprecht 106, pictured here.

Unlike most globular clusters, Ruprecht 106 may be what astronomers call a single population globular cluster. While the majority of stars in a globular cluster formed at approximately the same location and time, it turns out that almost all globular clusters contain at least two groups of stars with distinct chemical compositions. The newer stars will have a different chemical make-up that includes elements processed by their older, massive cluster companions. A tiny handful of globular clusters do not possess these multiple populations of stars, and Ruprecht 106 is a member of this enigmatic group.

Hubble captured this star-studded image using one of its most versatile instruments, the Advanced Camera for Surveys (ACS). Much like the stars in globular clusters, Hubble's instruments also have distinct generations: ACS is a third-generation instrument which replaced the original Faint Object Camera in 2002. Some of Hubble's other instruments have also gone through three iterations: The Wide Field Camera 3 replaced the Wide Field and Planetary Camera 2 (WFPC2) during the last servicing mission to Hubble. WFPC2 itself replaced the original Wide Field and Planetary Camera, which was installed on Hubble prior to its launch.

This high-tech tinkering in low Earth orbit has helped keep Hubble at the cutting edge of astronomy for more than three decades.

Astronauts on the space shuttle serviced Hubble in orbit a total of five times and were able to either upgrade aging equipment or replace instruments with newer, more capable versions.

Text credit: European Space Agency (ESA)
Image credit: ESA/Hubble & NASA, A. Dotter

Now let's have a look at the different types of STARS.

As we saw before, the energy for these stars comes from nuclear fusion whereby they fuse hydrogen atoms into helium atoms. Let's dig deeper into snow white and her three dwarfs.

Except, here snow white is the HR diagram, and the dwarf stars are yellow, red and white.

So, what really are DWARF STARS?

Dwarf stars, as can be seen by their name, are small stars that can go up to 20,000 times the luminosity of our sun and 20 times the mass of our sun. Our sun falls into this breed of stars.

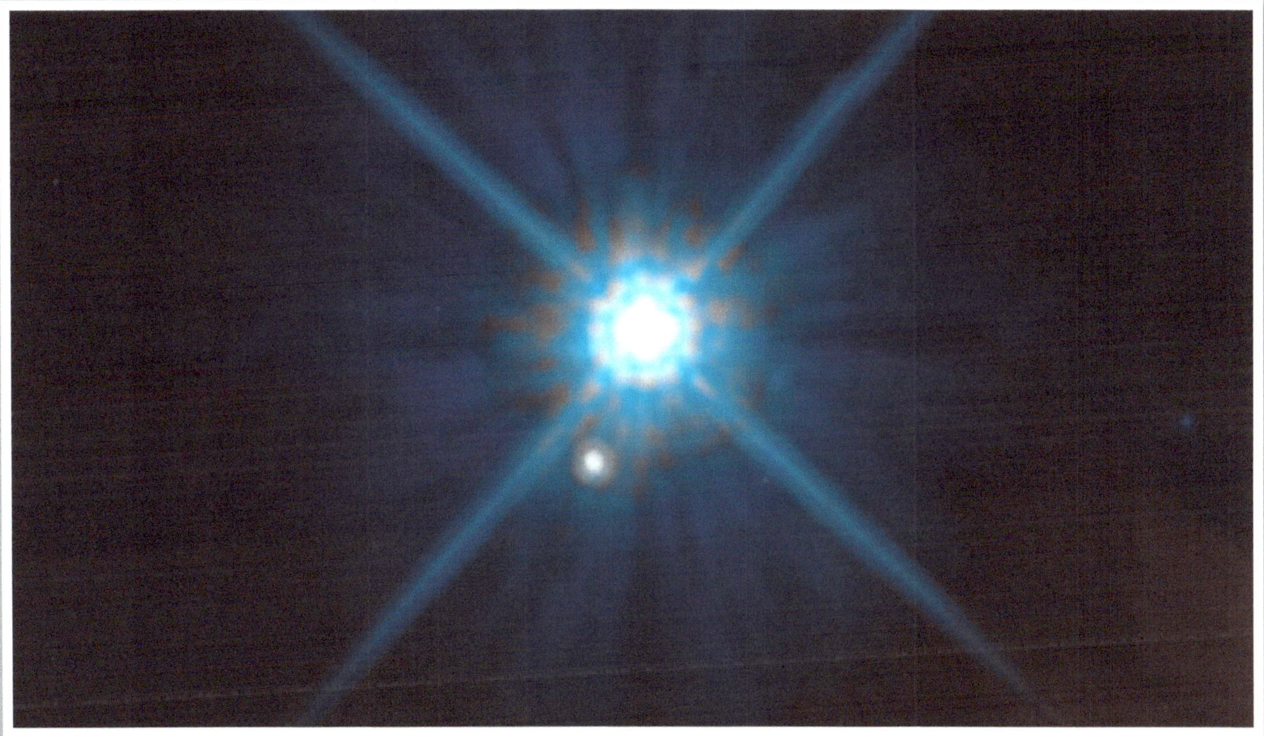

NASA, ESA, and K. Sahu (STScI)
WHITE DWARF STAR STEIN 2051 B

If we look further, we come across a certain species known as RED DWARFS.

What are their properties? Red dwarfs are cool, small, very faint, main sequence stars. Their surface temperature is less than 4000K and they're as common as cheese.

Even Proxima Centauri, the closest star to the sun (about 4.22 light years), is a red dwarf.

YELLOW DWARFS, on the other hand, are pretty basic. They're small and part of the main sequence.

Our sun is a yellow dwarf.

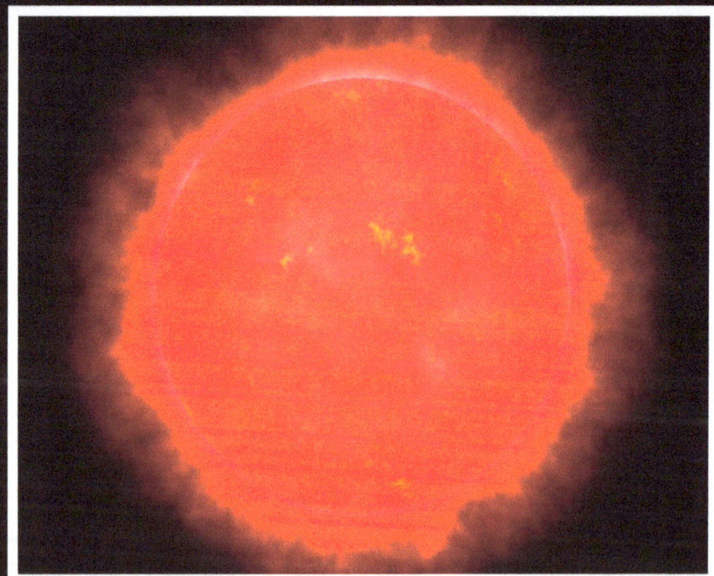

Red Dwarf

Alpha Centauri C, or better known as Proxima Centauri, is an example of a red dwarf. This star is notable because it is the closest star to our Sun at a distance of only about 4.2465 light-years away

https://theplanets.org/types-of-stars/red-dwarf-star/

Episode 2: Stars
https://www.theverge.com/2016/3/31/11338590/white-dwarf-star-oxygen-atmosphere-stellar-death

A white dwarf is a dense, small and hot star that is comprised mostly of CARBON. Yes, not hydrogen, not helium, but carbon.

Now let's come to the most interesting dwarf star — **THE WHITE DWARF**.

They are about the size of our own planet but enormously more massive. They will eventually lose all their heat and transform into dark, cold black dwarfs. Our sun will do so too.

Enough about dwarfs. Let's skip to the more fascinating part.

GIANTS.

The three types of giants we have in our universe are **RED, BLUE AND SUPERGIANTS**.

RED GIANTS are like the grandparents of stars. They're old. Having cooled, their diameter is 100 times larger than what it was originally. Their surface temperature is less than 6500 K. Although the name red giant, they are frequently orange in colour.

What are BLUE GIANTS? **BLUE GIANTS** are huge, hot and, as the name suggests, blue stars. These types of stars burn helium and are post-main sequence stars.

Okay guys, behold yourself, the boss is here.

SUPERGIANTS are the largest type of star ever known.

Some are even as large as our entire solar system.

However, these stars are rare.

But when they die, they go out with a bang.

Having caused a supernova, they become black holes when they die.

Wait, wait, wait! I know we just learnt about the boss of all stars, but something even more massive is still left.

A **NEUTRON STAR. A high mass star (10-40 solar masses), upon its death, leaves behind its massive core (1.4-3 solar masses).**

The gravity of this core will be so strong that the core won't be able to sustain itself. Hence, it will collapse with an enormous amount of force leading to electrons getting squeezed into protons. This will result in the formation of neutrons. What's left behind now is a small, but massive, object made of neutrons like one huge atomic nucleus as large as New York City.

A simulated view of a neutron star

You may complain about yourself gaining 7 kgs of weight over a month. You may avoid a burger because of its calories. But if you just take a teaspoon of a neutron star, you'll get a whopping ten million tons. You should probably consider getting a new weighing scale for that.

Alright. Now that we have a basic understanding of what the different types of stars are, let's plot them on the single most important diagram while studying stars.

THE HERTZSPRUNG-RUSSEL DIAGRAM

This diagram was developed in 1910 by astronomers Ejnar Hertzsprung and Henry Norris Russell, and it can be used to track how a star progresses through its life.

The temperature of a star is shown against its luminosity in this diagram. Now we have a basic idea of what the temperature of a star means. Let's take a look at luminosity. Luminosity is just another measure of brightness or the amount of light a star emits from its surface.

The Main Sequence of stars, which runs from top-left to bottom-right in the diagram, contains majority of stars, including our Sun.

As a result, they are located to the left and bottom of the main sequence. In most cases, stars will spend 90% of their lives on the main sequence before growing into a massive star for the final 10%. They will either go supernova or become a white dwarf after that.

Note that the x-axis of the HR Diagram (horizontal axis) can be written in a variety of ways, including the star's temperature (Kelvin), spectral class (OBAFGKM), or even colour.

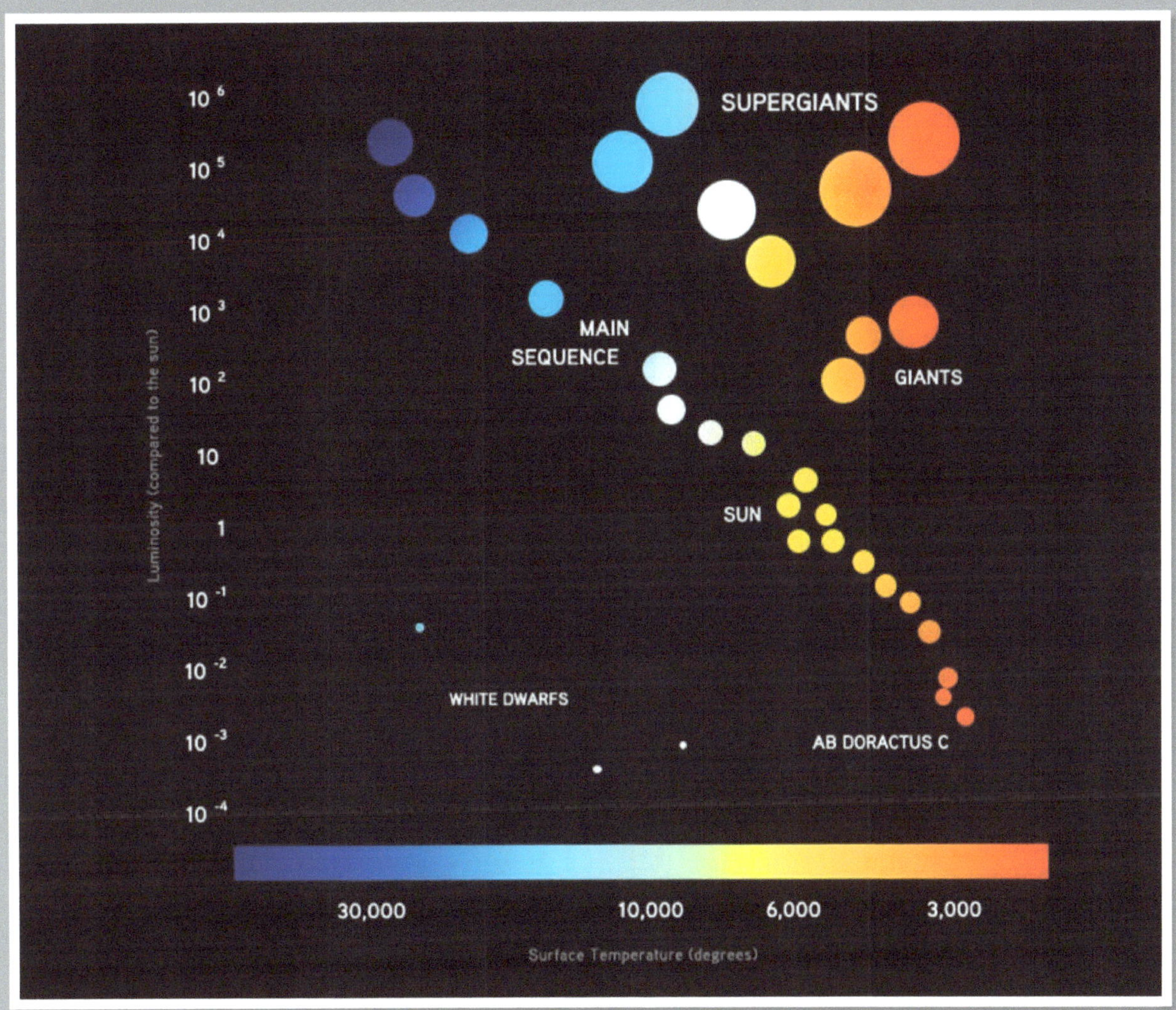

The Hertzsprung-Russel diagram

Now let's talk about our own star. You guessed it right,

THE SUN

Our Sun's birthday was around 4.6 billion years ago. Don't worry, it's not gonna die anytime soon unless we consider 5 billion years soon.

We feel extremely hot at 38° C, water boils at 100° C and the temperature of the sun's core is 15,000,000 °C. That's 15 million degrees celsius. So, if you plan on going there, don't forget your beach clothes and sun screen.

Meanwhile, the energy from this nuclear fusion is released in terms of light. This light immediately collides with a subatomic particle that absorbs the light and converts some of the energy into motion and re-emits the light with a little less energy. The light works its way to the surface of the sun in this manner and comes out as a photon. This process of light travelling from the core to the surface can take upto 200,000 years. That was all about our sun.

When a star's core runs out of hydrogen fuel, it can no longer maintain the outward radiation force that balances the gravitational force that pulls everything inwards.

As a result, the star will lose its hydrostatic equilibrium. The decrease of radiation pressure causes the core of the star to collapse. As a result of the core's collapse, the temperature of the inside rises.

The Sun is an example of a G-type main-sequence star (yellow dwarf). NASA Solar Dynamics Observatory.

The Sun comprises 91% Hydrogen, 8.9% Helium and 0.1% heavier elements like nitrogen and carbon.

The pressure at the sun's core is 260 billion times the earth's atmospheric pressure. At these extreme temperature and pressure conditions, the sun's core is fusing the hydrogen atoms into helium atoms. And it isn't just fusing 5 atoms a minute. The sun fuses 700 million tons of hydrogen to 695 million tons of helium every SECOND.

The lifecycle of a star (NASA and the Night Sky Network).

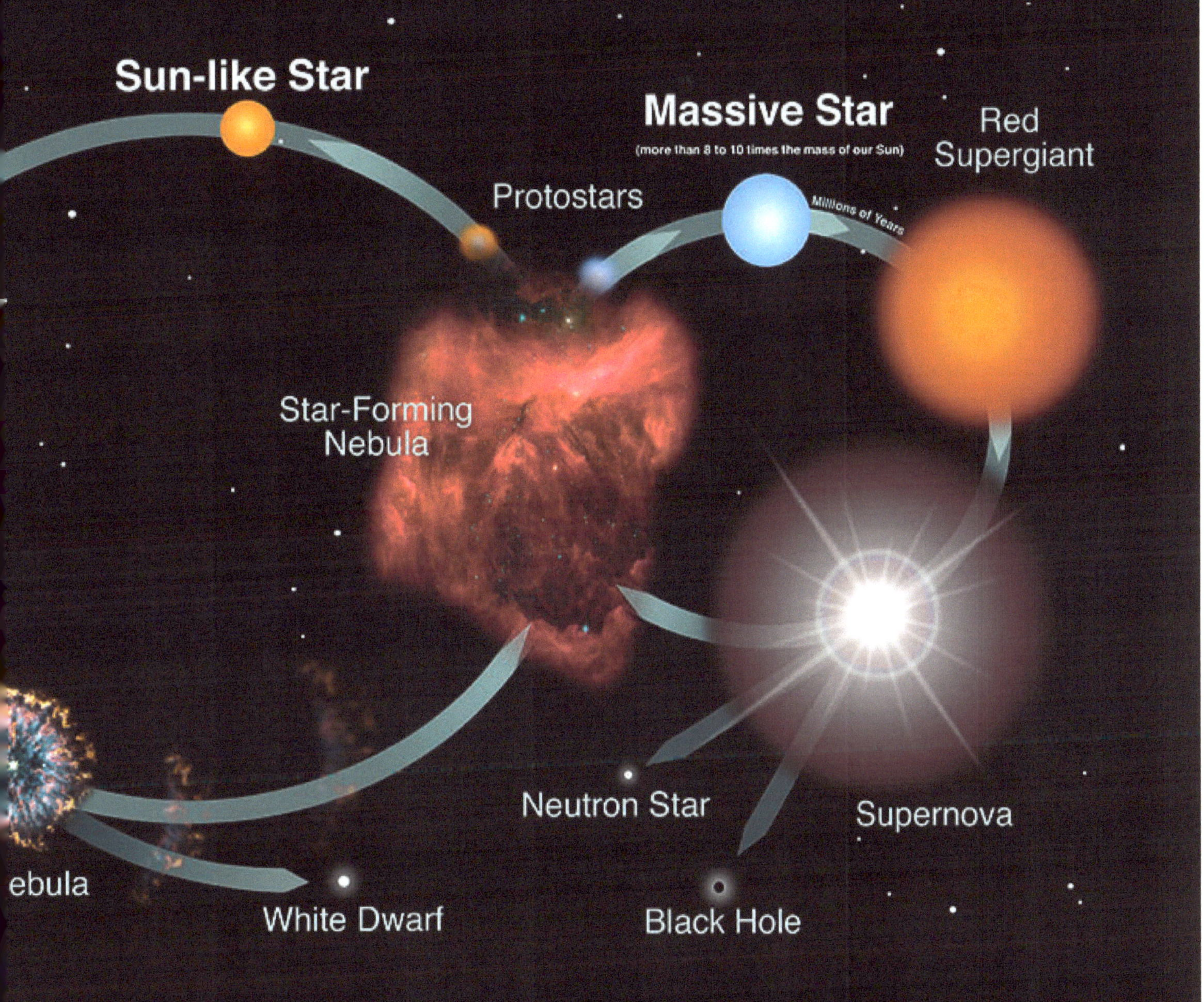

When a star that burns predominantly hydrogen undergoes such a collapse, the core of the star contracts until it reaches roughly 100 million degrees. At that temperature, the star's core can start to burn helium. As helium fuses to form carbon and other trace elements, it will become the star's primary source of energy.

However, if we look just outside the helium core, we can see that it is surrounded by a burning shell of hydrogen. As it spreads outward across the star, this hydrogen shell is steadily consuming more material.

Because helium burns hotter than its antecedent, the hydrogen core, it burns faster, resulting in a shorter phase of the star's existence. When the helium core is depleted, the core will collapse once more.

The temperature will rise once again when it falls. The star can burn carbon if the collapse causes a temperature of 600 million degrees. Shells of helium and hydrogen would surround such a core. After the carbon is burned out, the cycle can begin again, leading to the burning of neon, oxygen, and even silicon, with the core becoming heavier and heavier.

The star burns through the core more quickly as each layer becomes hotter than the last.

For example, a star's hydrogen may take billions of years to burn through, whereas a carbon core may just take hundreds of years to consume. By the time we reach silicon, a star's core may be consumed in about a day.

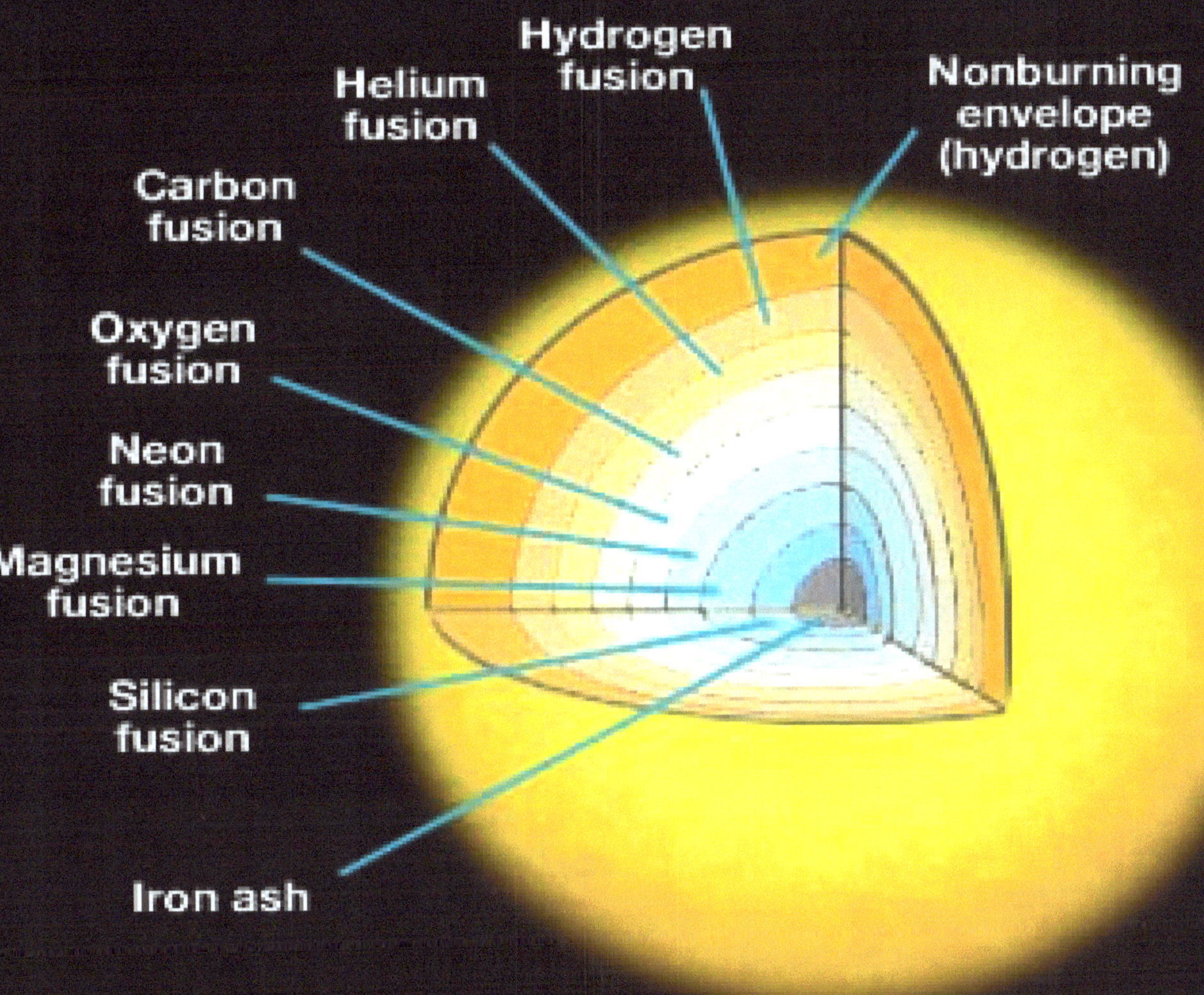

Artist's illustration of the core of a massive star just prior to a type II supernova explosion.

Because of the various levels visible above, these stars are frequently referred to as onion stars, as onions, like ogres, cakes, and parfaits, have layers. Returning to the stars. They do have layers. But it all comes to a halt with iron.

When learning about fusion and fission, we learnt that iron is the most closely connected atom, and that breaking iron apart or smashing two iron nuclei together yields no more energy. As a result, the star has run out of fuel to burn through. The fusion in the core has to stop and the star will die.

But from the remnants of stars, heavier elements are cast into the universe, and it is this star dust that form the seedlings of life itself.

EPISODE 3:
GALAXIES

I know we've already spoken A LOT about big stuff. Really huge stuff. However now we're going to talk about even more enormous stuff. GALAXIES.

So, what are GALAXIES really?

Galaxies are basically concentrations of stars, dust and gas all held together by gravity.

Credit: ESA/Hubble & NASA, RELICS

**https://en.wikipedia.org/wiki/
Spiral_galaxy**

Furthermore, a galaxy is a gravitationally bound system that consists of stars, stellar remnants, interstellar gas, dust, and dark matter.

Galaxies range in size from dwarfs with a few hundred million stars to giants with a hundred trillion stars, all orbiting their galaxy's centre of mass.

Most galaxies are between 1.000 to 100.000 parsecs / 3.000 to 300.000 light-years in diameter.

Recent calculations of the observable universe suggest that there are over 2 TRILLION GALAXIES that contain an estimated 1×10^{24} stars.

Galaxies are classified as ELLIPTICAL, SPIRAL, OR IRREGULAR based on how they look.

We've seen **SPIRAL GALAXIES**, and we live in one!

Broad, flat revolving discs of stars, gas, and dust distinguish them.

These spirals are also available in a range of varieties.

Grand Design spirals have arms that are majestic and magnificent.

These spiral arms extend from the galaxy's very centre to the observable edge of the galaxy. Other spiral galaxies have choppy or scattered arms, which are known as flocculent spirals and resemble cotton tufts.

Now don't start craving cotton candy just because I mentioned cotton tufts.

Credit: ESA/Hubble & NASA, RELICS

ELLIPTICAL GALAXIES, as the name implies, are elliptical. Sort of.

Some are virtually spherical, massive cotton balls filled with billions of stars. Others are elongated, resembling cigars or American footballs in shape.

They don't have any particular overall structure as spirals do, and they're puffy.

They range in size from dwarf ellipticals — barely a few thousand light years across — to giants that dwarf our own Milky Way.

https://en.wikipedia.org/wiki/Elliptical_galaxy

www.eso.org

ESO.org The irregular galaxy NGC 1427A | ESO

IRREGULAR GALAXIES, on the other hand, have no definite shape but are in constant motion like all other galaxies. They don't appear to have a nuclear bulge or any evidence of spiral arms, giving them a chaotic look. Irregular galaxies are typically small, weighing roughly a tenth of the mass of our Milky Way.

That was it from these different flavours of galaxies. Let's move on to something more interesting now.

GALAXY COLLISIONS

Now that we know what galaxies are, let's see what happens when galaxies COLLIDE. When galaxies clash, it's like a **cosmic train crash**.

Galaxies are massive objects containing hundreds of billions of stars that generate a lot of gravity. Two galaxies can lure each other in and collide if they get near enough together. A galactic collision is a strange occurrence.

Tidal effects* can be severe in the early phases of a collision.

Despite the fact that collision speeds can reach hundreds of kilometres per second, the event takes hundreds of millions of years to play out — remember, we're talking about tens of thousands of light years here.

Stars on the side of the galaxy closest to the other are more powerfully drawn toward it than stars further away, causing galaxies to stretch and long tendrils of stars and gas to be drawn out.

Collisions are usually not head-on, but rather sideswipes, resulting in some sideways motion.

The tidal streamer can then become curled, forming a long, elegant arc. In this stage, colliding galaxies produce a wide variety of odd and magnificent formations.

But this isn't the same as two trucks colliding.

Because stars are so little in comparison to the space between them, even if hundreds of billions of them are involved, there's a significant chance that no two will ever physically collide with each other!

Space is odd.

Gas clouds, on the other hand, are enormous and do collide.

They collide, implode, and produce stars at an incredible speed. As stars form and light up the hydrogen clouds around them, colliding galaxies can glow pink and blue.

Occasionally, two galaxies collide at such a high speed that they pass directly through each other! Even then, they'll usually slow down, stop, and then recollide. They eventually combine to form elliptical galaxies.

The galaxies NGC 5394 (the smaller one, on the right) and NGC 5395 (the larger one, on the left) are in the middle of colliding over the span of millions of years.

NSF's National Optical-Infrared Astronomy Research Laboratory/Gemini Observatory/AURA

MILKY WAY GALAXY

Let's stop for a second and learn about our own galaxy for a minute.

It's a gigantic pinwheel of stars and gas whirling within a massive cloud of gas and dust bound together by gravity. The Milky Way galaxy is a large spiral galaxy.

Our Sun, which is a star, is part of the Milky Way Galaxy, as is Earth and all of the planets that orbit it.

When you go stargazing, all of the stars you see are part of the Milky Way galaxy.

The Milky Way is named for the fact that it appears in the sky as a milky stretch of light when viewed from a very dark location.

We get angry if we reach a place 10 minutes late due to traffic. We get annoyed if our flight gets delayed by 2 hours. The Milky Way galaxy's disc is around 100,000 light years across.

https://solarsystem.nasa.gov/resources/285/the-milky-way-galaxy/

A Roadmap to the Milky Way

(artist's concept)

ASA / JPL-Caltech / R. Hurt (SSC-Caltech)　ssc2008-10a

This means that it will take light (whose speed is 3×10^8 m/s) 100,000 YEARS to cover the milky way from top to bottom. I can say that that's a pretty huge distance.

ANDROMEDA GALAXY

Just how being at home all the time due to the covid-19 pandemic had annoyed us out of our minds, let's also leave home (our Milky Way galaxy) and explore beyond.

The Andromeda galaxy, unlike most other galaxies that are moving away from Earth due to the expansion of the universe, is speeding towards us at a whacking speed of roughly 110 kilometres per second.

So, what's the nearest major galaxy next to earth?

It's the ANDROMEDA GALAXY. This galaxy has a diameter of roughly 220,000 light years. The enormous spiral galaxy, which is 2.5 million light years away, includes about one trillion stars.

Andromeda has had a rough past. It wasn't always like this. Andromeda has grown in size and swallowed a number of other galaxies during the course of its existence.

Assume you've found a way to go to Andromeda.

What do you think you'd see? What might the galaxy's appearance be like?

Andromeda, like the Milky Way, is a spiral galaxy.

When Andromeda Collides with The Milky Way

As we saw earlier,

Andromeda is speeding towards us at the furious rate of 110 kilometres per second.

+ **Andromeda will eventually collide with our Milky Way in a massive galactic collision four billion years from now, but what if they collided tomorrow?**

+ **Will all of the stars and planets crash?**

+ **Will our solar system survive this tremendous collision?**

+ **How will this celestial light display seem from Earth?**

If the Milky Way and the Andromeda galaxy collided, this is what would happen: In the observable universe, there are at least 100 billion galaxies colliding with one another. Larger galaxies combine every 9 billion years, while smaller galaxies collide even more frequently.

Andromeda is
predicted to
be around 25%
more luminous
than our own
galaxy.

https://www.jpl.nasa.gov/images/pia15416-
andromeda

What happens now?

To begin with, there would be no Milky Way and no Andromeda Galaxy. The two spiral galaxies would merge to form a new galaxy called **MILKOMEDA**, which would be elliptical in shape.

After Andromeda enters Milky Way's personal space, the Sun and our whole solar system would be forced to the outskirts of the newly created galaxy 26,000 light years away from home.

What about all the stars in both galaxies? Will they crash into each other?

Some of them might, but despite the fact that the Milky Way has over 250 billion stars and Andromeda has over a trillion, they are separated by light years of empty space.

Many of them are unlikely to crash. Instead, they simply spread into other orbits.

I almost forgot to mention that this celebration is also attended by a third galaxy. M33, often known as the TRIANGULUM GALAXY, is a tiny satellite galaxy of Andromeda.

This is turning up to be quite a mash-up.!

https://en.wikipedia.org/wiki/Andromeda%E2%80%93Milky_Way_collision#/media/
File:Andromeda_Collides_Milky_Way.jpg

Based on data from the **Hubble Space Telescope**, the Milky Way Galaxy (pictured right-centre) and Andromeda Galaxy (left-centre) are predicted to distort each other with tidal pull in 3.75 billion years, as shown in this illustration.

For us, Earthlings, this would be breath taking. The new giant galaxy's brilliant core would light up the night sky. **The two supermassive black holes at the centre of these galaxies would eventually come so close to merging that they would fuse.** All of the gas swallowed by this massive black hole would condense into a dazzling quasar* at the galaxy's centre. Another fantastic performance for us here on Earth.

Unfortunately, our planet will not live long enough to witness any of it. The Sun would have already grown into a red giant and gobbled up the entire Earth by the time Andromeda collides with the Milky Way, so Earth will have other issues.

Nonetheless, the additional flood of dust should encourage star formation in the new "Milkomeda" galaxy, and the Earthless sun may probably depart the Milky Way for forever.

Conclusion

The expanse of the universe, with its unimaginable number of galaxies, is probably the most compelling reason for life to exist outside of our Solar System.

Innumerable stars can be found in all these galaxies, and these stars may have innumerable exoplanets orbiting them. Life should almost certainly exist somewhere else as well, at least from this perspective. Talking about exoplanets, let's now move on to how PLANETS work.

FUN FACTS ABOUT GALAXIES

1. **MACS 2129-1, often known as the Zombie Galaxy**, is a disk-shaped galaxy that rotates twice as quickly as our Milky Way. For the past 10 billion years, this galaxy hasn't produced any stars.

2. **W2246-0526 is one of the brightest galaxies in the cosmos**. This galaxy is simultaneously "eating" the mass of three neighbouring galaxies.

3. **Little Cub is a dormant dwarf galaxy** that has existed since the Big Bang. It could still include molecules that haven't changed since the universe's rapid expansion about 13.7 billion years ago.

EPISODE 4:
PLANETS

Now that we've studied stars in detail already, let's look at the massive objects that orbit around these stars. We can say that the planets are as obsessed with their stars (considering the fact that they circle these stars throughout their life

Before coming to the famous eight planets (without Pluto, sadly) of our solar system, let's quickly understand how planets are formed in the first place.

If you're yet wondering, but what essentially ARE planets? Here's the simplest answer:

✦ **A planet is a huge object that orbits a star, such as Venus or Earth.**

✦ **Planets are far smaller than stars and emit no light.**

✦ **Jupiter is the Solar System's largest planet. Planets have the shape of a slightly squished ball (referred to as a spheroid).**

There are innumerable galaxies in the universe, each with a large number of stars. Definitely more than the number of hairs on your head. There are planets orbiting some of those stars, just as there are planets orbiting our own Sun.

A theory called **'THE NEBULAR HYPOTHESIS'** is the most widely accepted theory for **how our solar system may have formed** when it was first formed. So, let's get straight into it.

According to this hypothesis, our Solar System's Sun and all of its planets originated as a massive cloud of molecular gas and dust. Something happened about 4.57 billion years ago that caused the cloud to collapse. This could have been caused by a passing star or supernova shock waves, but the end result was a gravitational collapse at the cloud's centre.

Mercury

Venus

Mars

Jupiter

Uranus

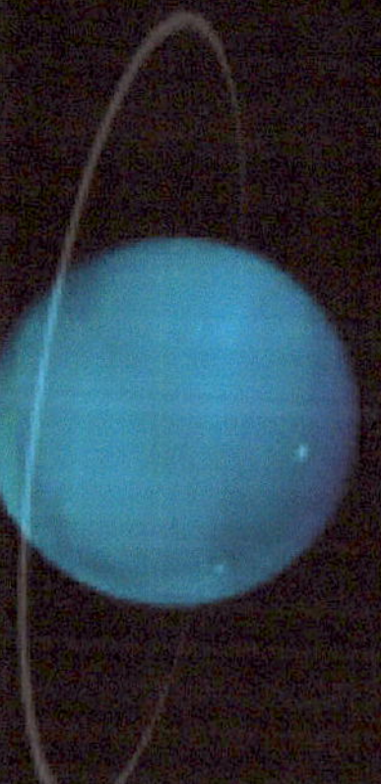

Neptune

Earth
S-NPP VIIRS
2015

Saturn
Cassini-Huygen
2000 (planet)
2007 (rings)

Pluto
New Horizons
2015

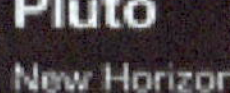

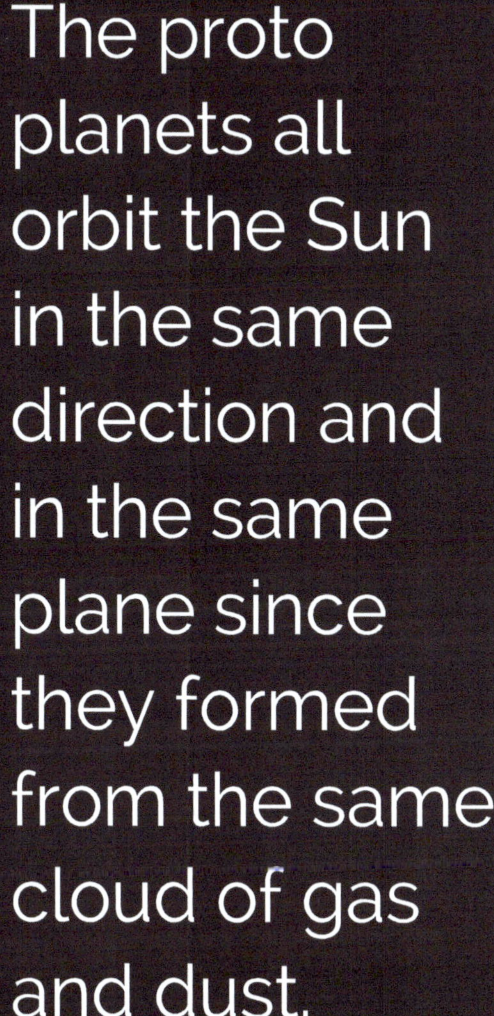

The proto planets all orbit the Sun in the same direction and in the same plane since they formed from the same cloud of gas and dust.

Pockets of dust and gas began to gather in denser areas as a result of the collapse. As more matter was drawn in by the denser regions, conservation of momentum forced it to begin rotating, while increasing pressure caused it to heat up. Outside the star, matter was clumping together to form protoplanetary clumps of gas, dust, and rock. As material was trapped in their gravitational fields, these protoplanets continued to grow.

Only metals and silicates could live in solid form closer to the Sun due

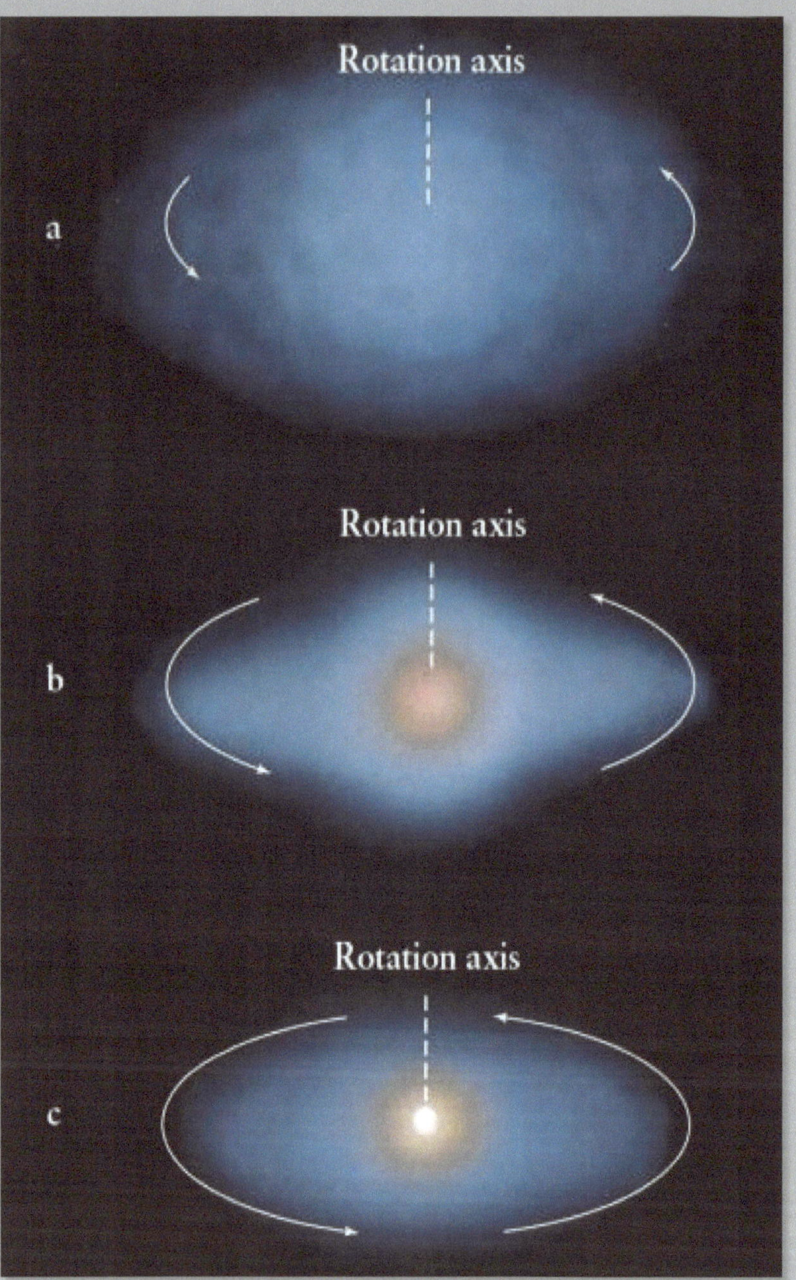

Spinning cloud flattening into a disk and condensing into a star and planets

Contracting nebular cloud increasing its rate of rotation

to their higher boiling points, and these would eventually become the terrestrial planets of **Mercury, Venus, Earth, and Mars**. The terrestrial planets could not develop very large because metallic elements made up such a tiny part of the solar nebula.

The large planets **(Jupiter, Saturn, Uranus, and Neptune)**, on the other hand, developed beyond Mars where material is cool enough for volatile icy compounds to solidify. (i.e. the Frost Line).

These planets were huge enough to collect vast atmospheres of hydrogen and helium because the ices that built them were more abundant than the metals and silicates that formed the terrestrial inner planets.

So, in this game of rock, paper and scissors, an extra player is added to the game — ice. And of course, as we saw, ice beats rock in terms of abundance but rock beats ice in terms of proximity.

It explains why planets and the stars they circle usually spin and lie in the same plane. It also explains the planets' order, with the rocky planets being closest to the Sun and the gas giants being further away.

Within 50 million years, the pressure and density of hydrogen in the protostar's core had increased to the point where thermonuclear fusion could commence. I feel a star coming. Temperature, reaction rate, pressure, and density all increased until they reached hydrostatic equilibrium.

The Sun became a main-sequence star at this point.

Despite the fact that the nebular theory cannot be directly tested, it provides a useful description of **how our solar system developed.**

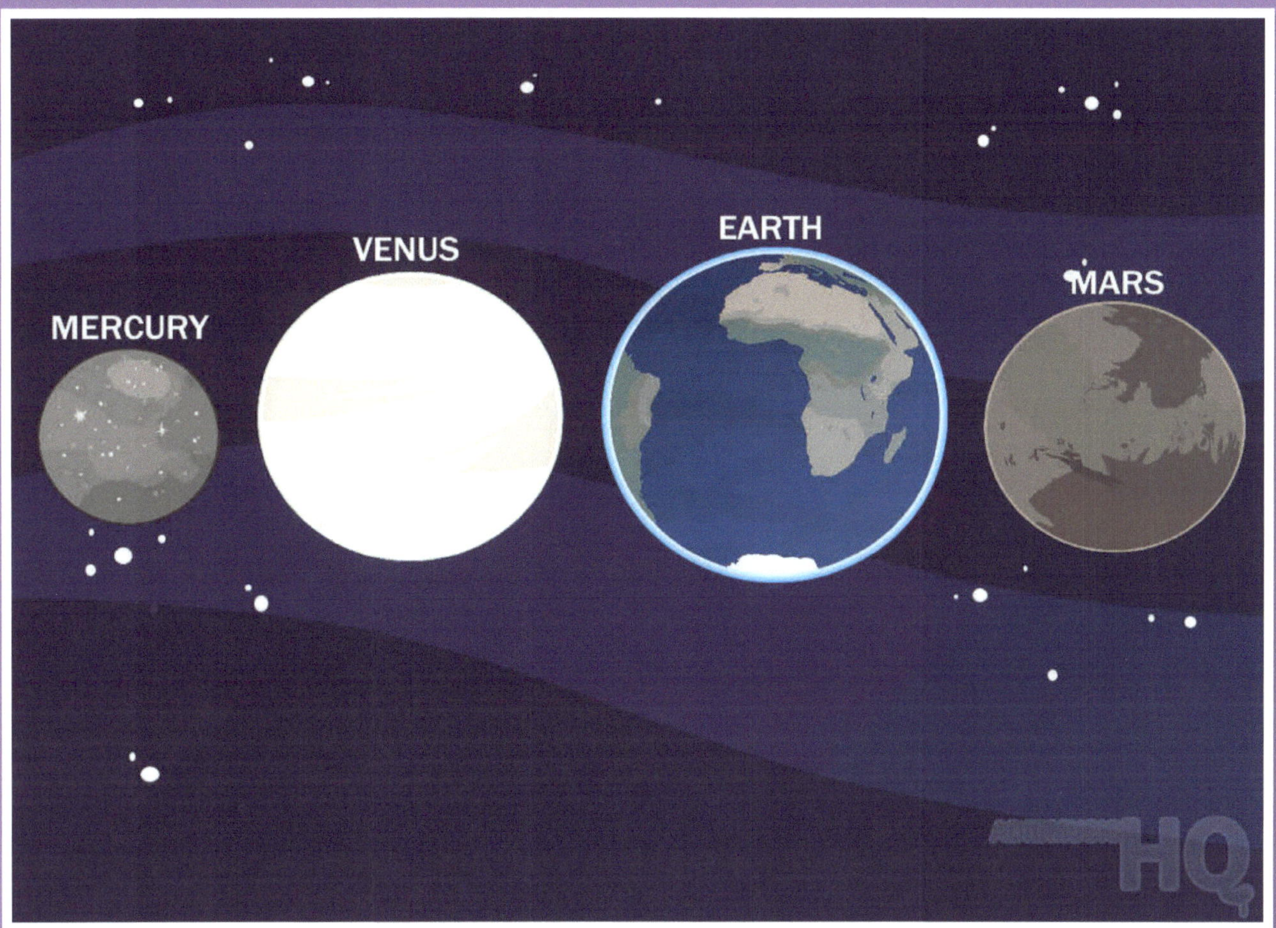

So now we know that this order of planets isn't something random that came in God's dream.

It has practical sense.

Now let's put the put the idiom "home sweet home" to use and study about our 8 planets.

There are eight planets that orbit our sun.

The planets are classified into **TERRESTRIAL and JOVIAN** on the basis of their composition.

MERCURY, VENUS, EARTH, AND MARS are all terrestrial planets that are mostly composed of rocky material.

They have solid surfaces, no ring systems, few or no moons, and are relatively tiny.

Venus rotates in a direction opposite to the other planets.

This means that on Venus the Sun rises in the west and sets in the east.

MERCURY, the smallest and nearest planet to the sun, has the shortest orbit in the solar system, taking around three Earth months.

VENUS is the hottest planet in the solar system, with temperatures reaching 867 degrees Fahrenheit due to a carbon dioxide atmosphere and widespread lava flows.

There are three theories to explain this.

1. An explanation for the backward, or retrograde, rotation is not certain. A long-held theory is that Venus once rotated as the other planets do but was struck billions of years ago by a planet-size object. **The impact and its aftermath caused the rotation to change directions or flipped the planetary axis.**

Now let's look at the other two. theories

2. Current theory holds that Venus initially spun in the same direction as most other planets and, in a way, still does it simply flipped its axis 180 degrees at some point. In other words, **it spins in the same direction it always has, just upside down, so that looking at it from other planets makes the spin seem backward.** Scientists have argued that the sun's gravitational pull on the planet's very dense atmosphere could have caused strong atmospheric tides. Such tides, combined with friction between Venus's mantle and core, could have caused the flip in the first place.

3. A more recent idea, put forth in the journal Nature in 2001, suggests that an initial counter-clockwise rotation was an unstable state for various complex reasons and that **the planet slowed down and slipped into a more stable state of clockwise rotation.**

This image compares the retrograde rotation of Venus to the direct rotation of Earth. Windows original, courtesy of Nate Proulx.

EARTH, which is located next to this land of fire, is a world of water. The water systems on our planet contribute to the creation of the universe's only known environment capable of supporting life.

Now let's have a look at the one planet films and TV are obsessed with — **MARS.** Mars, the last of the terrestrial planets, might just have sustained life as recently as 3.7 billion years ago, when it had a watery surface and a moist atmosphere.

The Jovian planets of the outer solar system lie beyond the four Terrestrial planets of the inner solar system.

The gas giants Jupiter and Saturn, as well as the ice giants Uranus and Neptune, make up the Jovian planets. The ice giants comprise rock, ice, and a liquid mixture of water, methane, and ammonia, whereas the gas giants are mostly formed of helium and hydrogen. In fact, one of Saturn's moons, Enceladus, has oceans of methane.

Each of the four Jovian planets has several moons, ring systems, no solid surface, and is enormous.

JUPITER is the largest Jovian and the largest planet in the solar system.

HOW ABOUT LANDING ON JUPITER?

Landing on a new planet is the finest way to begin your exploration. That is the reason why people have sent spacecraft to places like the **Moon, Venus, Mars, Titan, and Saturn's moon.**

Jupiter, the largest planet in the solar system, is a mysterious place. Through images taken by telescopes and the Juno probe, people on Earth can admire Jupiter's splendour. On the other hand, we might never be able to see what is hidden beneath the planet's thick, whirling clouds.

The main gases that make up Jupiter are hydrogen and helium.

Attempting to land there would be similar to attempting to do so here on a cloud.

We will never fully comprehend some regions of the solar system, though. Jupiter is one of them.

What if one still attempted to land on Jupiter is as follows:

It's vital to note that the Lunar Lander is the focus of the descent's first half. The Lunar Lander is really less fragile than, say, NASA's Orion spacecraft.

Since Jupiter has an atmosphere, the Lunar Lander would not be appropriate for a mission to land there. However, even the most durable spacecraft would not last long in Jupiter, making the Lunar Lander the best option for this fictitious scenario.

First things first, there is no oxygen in Jupiter's atmosphere. Therefore, be careful to pack enough air for yourself. The sweltering heat are the next issue. So, make sure to bring a fan.

You are now prepared to embark on an incredible quest.

Here's how many Earths you could stack from Jupiter's centre for scale. As you approach the top of the atmosphere, you'll be travelling at 110,000 mph due to Jupiter's gravity.

But hold your breath. You'll hit the denser atmosphere below quickly, which will hit you like a wall. But it won't be enough to deter you.

After about 3 minutes, you'll be 155 miles below the clouds. You'll feel the full force of Jupiter's rotation here.

A day lasts approximately 9.5 Earth hours. This generates powerful winds that can travel at speeds of more than 300 miles per hour around the planet.

Jupiter is the planet with the fastest rotation in our solar system.

The pressure is nearly 100 times greater than at the Earth's surface. And because you won't be able to see anything, you'll have to rely on tools to explore your surroundings.

The pressure is 1,150 times higher 430 miles down. You might be able to survive down here if you were in a spacecraft built like the Trieste submarine, the world's deepest diving submarine. Any deeper, the pressure and temperature will be too high for a spacecraft to withstand.

But suppose you could find a way to descend even further. You will solve some of Jupiter's most enigmatic mysteries. **Unfortunately, you won't be able to tell anyone. Because Jupiter's dense atmosphere absorbs radio waves, you will be cut off from the outside world.**

Jupiter as seen by the space probe *Cassini*

The temperature is 6,100 degrees Fahrenheit once you've travelled 2,500 miles down. That's hot enough to melt tungsten, the metal with the highest melting point in the Universe.

You'll have been falling for at least 12 hours at this point. And you won't even be halfway through.

Jupiter's innermost layer is reached after 13,000 miles.

The pressure is 2 million times greater than on Earth's surface. And the temperature is hotter than the surface of the sun.

These conditions are so extreme that they alter the chemistry of the hydrogen around you. Hydrogen

https://en.wikipedia.org/wiki/Exploration_of_Jupiter#/media/File:Portrait_of_Jupiter_from_Cassini.jpg

molecules are forced so close together that their electrons break free, resulting in metallic hydrogen, an unusual substance. Metallic hydrogen is extremely reflective.

So, trying to see down here with lights would be impossible.

It's also as dense as a rock. As you go deeper, the buoyancy force of the metallic hydrogen compensates for gravity's downward pull.

Like a yo-yo, buoyancy will eventually propel you back up until gravity pulls you back down. And when those two forces balance, you'll be stuck in mid-Jupiter, unable to move up or down and with no way out!

To summarise, attempting to land on Jupiter is a bad idea. We may never know what lies beneath those magnificent clouds. From afar, we can still study and admire this mysterious planet.

Next is **SATURN, the solar system's second largest planet**.

Its distinctive rings are nearly a kilometre thick and wide enough to fit between Earth and the moon.

Saturn is the sixth planet from the sun.
Credits: NASA
https://www.nasa.gov/audience/forstudents/k-4/stories/nasa-knows/ring-a-round-the-saturn.html

URANUS and NEPTUNE, the icy giants, come after Saturn.

Uranus, the somewhat larger of the ice giants, is known for spinning on its side.

Uranus, the somewhat larger of the ice giants, is known for spinning on its side.

Next in line is Neptune, the solar system's farthest planet and one of the coldest.

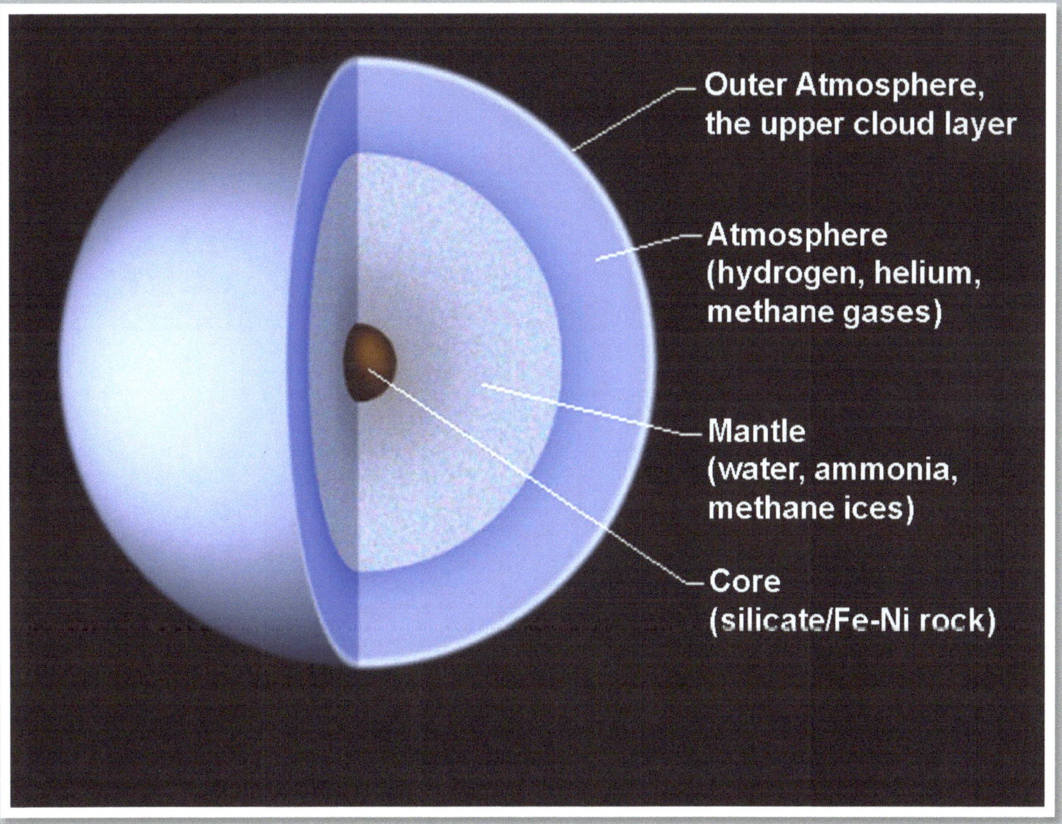

http://abyss.uoregon.edu/~js/ast121/lectures/lec20.html

Uranus (L) and Neptune (R) as seen from Voyager
https://www.theverge.com/2017/6/16/15810926/nasa-uranus-neptune-mission-voyager-2-spacecraft

Neptune has eight moons, six of which were found by Voyager.

A day on Neptune is 16 hours and 6.7 minutes long. Neptune was discovered on September 23, 1846 by Johann Gottfried Galle, of the Berlin Observatory, and Louis d'Arrest, an astronomy student, through mathematical predictions made by Urbain Jean Joseph Le Verrier.

https://solarsystem.nasa.gov/
news/2219/new-study-finds-unexpected-
temperature-changes-on-neptune/

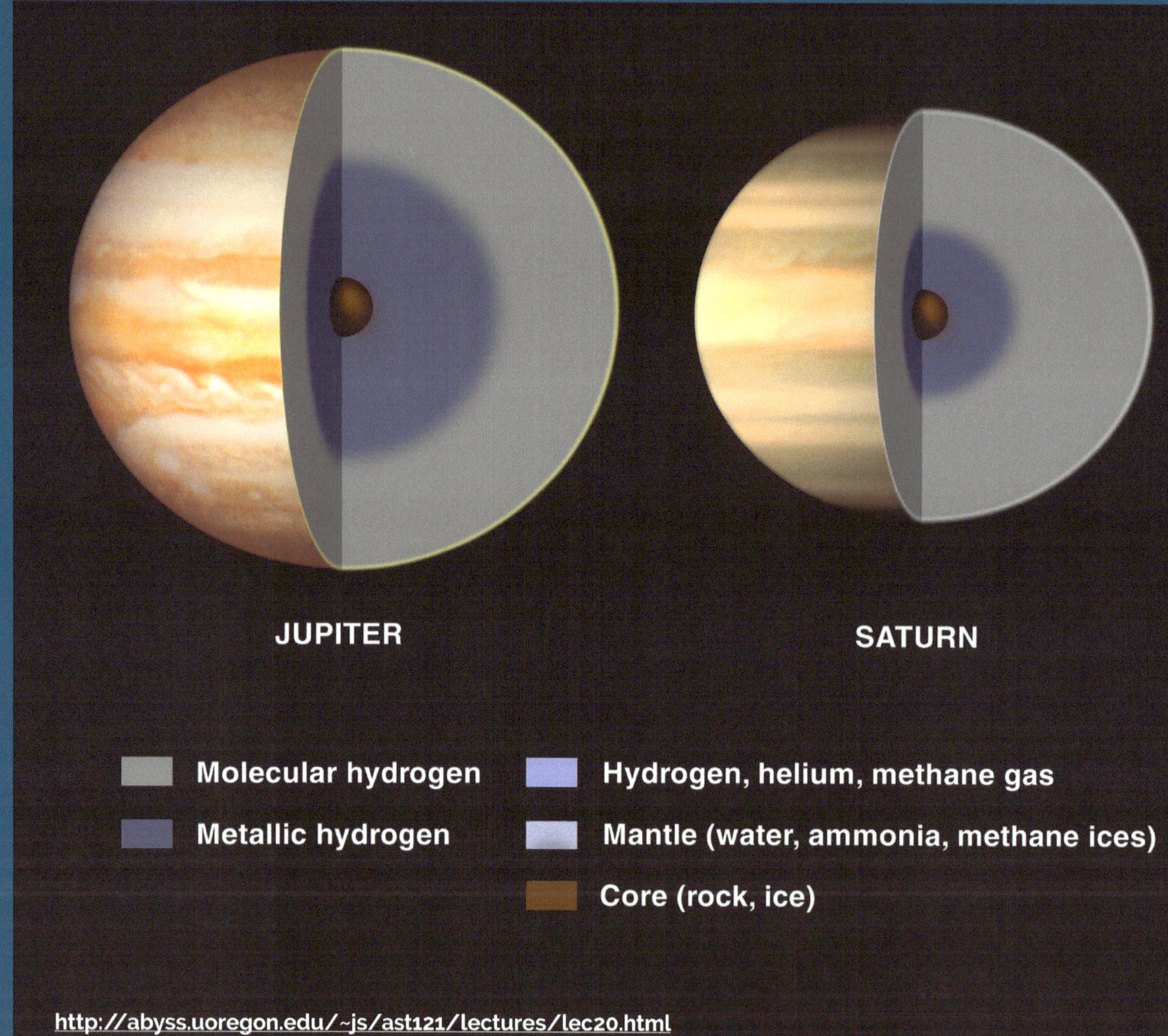
JUPITER
SATURN
Molecular hydrogen
Metallic hydrogen
Hydrogen, helium, methane gas
Mantle (water, ammonia, methane ices)
Core (rock, ice)
http://abyss.uoregon.edu/~js/ast121/lectures/lec20.html

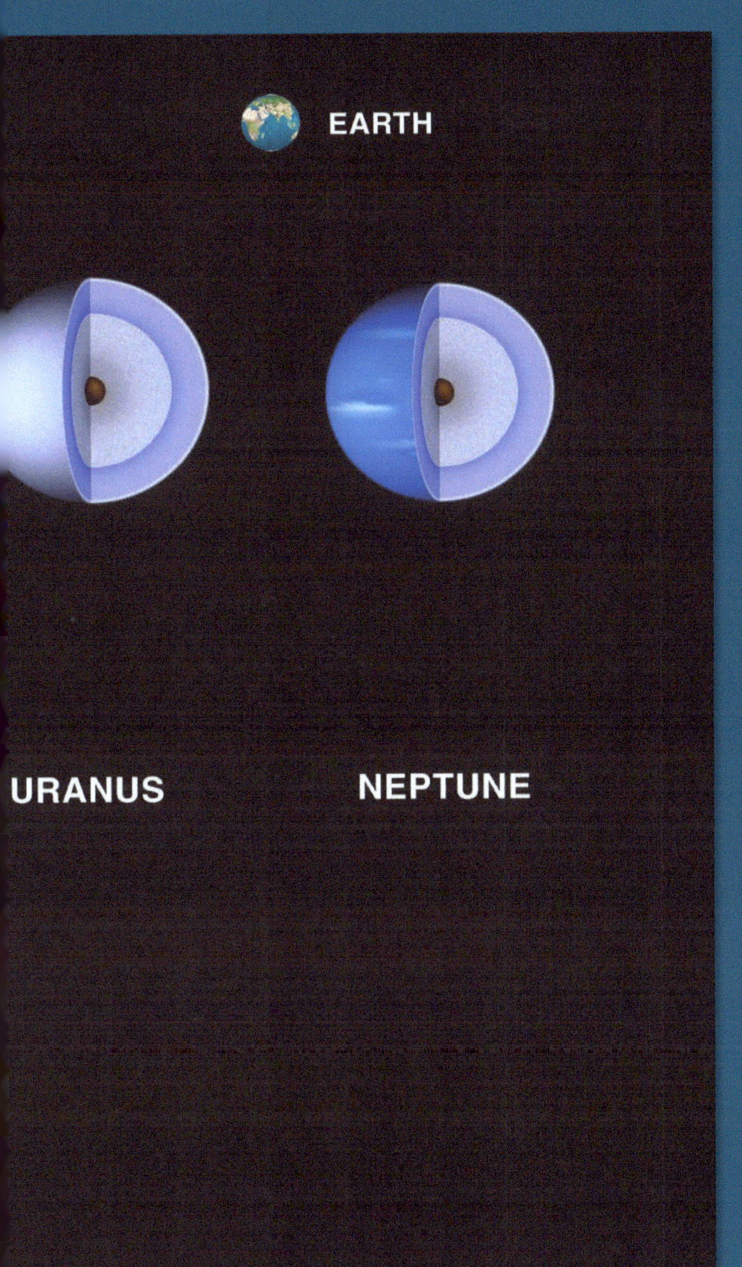

This was the list of all the planets you recited in order when you were in the 3rd grade. Now let's think out of the box and look at some planets that we aren't quite familiar about. **EXOPLANETS.**

EXOPLANETS — PLANETS BEYOND OUR SOLAR SYSTEM

They're buried in the final frontier, numerous worlds strewn across many galaxies challenging the idea that we're the only ones in the cosmos.

Exoplanets are planets that orbit the sun but are not part of our solar system.

A planet beyond our solar system, or one that does not orbit our sun, is referred to as an exoplanet. The majority orbit other stars, while some are lonely wanderers in the interstellar vacuum. Exoplanets

are exceedingly far away due to the fact that they orbit other stars. Most are too far away to even consider sending space probes to investigate.

So, why are we interested in EXOPLANETS?

Our desire to discover worlds with life is, of course, a major factor.

We'd like to learn more about what makes a planet habitable, or a place with all the necessary components and conditions for life.

How does it happen?

How common is it?

We believe that the planets that most resemble Earth in terms of size and composition, as

well as being at the correct distance from their stars to have liquid water on their surfaces, are the most life-ready.

These features are difficult to observe from such a distance, but we're working on it. Exoplanets also teach us a variety of other things. Understanding the history of our own planet family, including Earth, is enhanced by studying other planetary systems. In the 1990s, scientists discovered the first exoplanets. But now we're discovering tens of thousands more.

However, the discovery of an exoplanet orbiting a sun-like star was only confirmed in 1995. Since then, nearly 4000 exoplanets have been confirmed using a variety of methods.

For thousands of years, scientists have speculated the existence of worlds outside our solar system.

The quest for exoplanets continues, not just to build a database of known worlds, but also to see if life could exist elsewhere in the cosmos.

TYPES OF EXOPLANET

NEPTUNE-LIKE
Similar to Uranus or Neptune.

SUPER-EARTH
Bigger than the Earth, but lighter than Neptune.

GAS GIANTS
Similar in size compared to Saturn and Jupiter.

TERRESTRIAL
Rocky and almost the same size as the Earth.

FUN FACTS ABOUT PLANETS:

1. The name "planet" is from the Greek word **πλανήτης (planetes),** meaning **"wanderers",** or "things that move".

2. **The winds on Neptune are supersonic**. While hurricanes create panic on Earth, the power of these storms is nothing compared to what you'd see on Neptune. According to NASA, winds may reach speeds of over 1,100 miles per hour (1,770 kilometres per hour). To put it in perspective, that's quicker than the speed of sound at sea level on Earth. It's a mystery why Neptune is so windy, especially given how little the Sun's heat is at such a great distance.

3. **Venus doesn't have any moons**, and we aren't sure why. Both Mercury and Venus have no moons, which is surprising given the fact that there are dozens of others scattered around the Solar System.

4. **Saturn, for example, contains nearly 60 moons**. And other moons are nothing more than captured asteroids, as is the case with Mars' two moons, for example. So, what distinguishes these planets? No one is really sure why Venus doesn't, but there is at least one stream of research that suggests it could have had one in the past.

5. **The Moon takes about one month to orbit Earth (27.3 days to complete a revolution, but 29.5 days to change from New Moon to New Moon).**

Moon's orbit is slightly tilted from the orbit of the earth around the sun. The plane of the lunar orbit is inclined to the ecliptic by about 5.1°.

6. **The time it takes for the Moon to *rotate* once on its axis is equal to the time it takes for the Moon to *orbit* once around Earth.** This keeps the same side of the Moon facing towards Earth throughout the month. If the Moon did not rotate on its axis at all, or if it rotated at any other rate, then we would see different parts of the Moon throughout the month.

Enough about all these massive objects now.

Let's start talking about **COMETS AND METEORS**.

EPISODE 5:
COMETS AND METEORS

"If this comet makes impact, it will have the power of a billion Hiroshima bombs." This is a quote from the movie 'Don't Look Up' that aired on Netflix in December 2021. Starring Jennifer Lawrence and Leonardo DiCaprio, this film was a journey of two astronomers discovering a comet that was bigger than the asteroid that killed the dinosaurs. This comet was hurling towards earth at gigantic speeds, and they had precisely six months until the comet tore through earth's atmosphere and brought all life to extinction.

So, let's talk more about these comets.

People used to be both awestruck and scared by comets, which they saw as long-haired stars that came in the sky unexpectedly and unpredictably. For millennia, Chinese astronomers kept meticulous records, including drawings of different types of comet tails, dates of cometary appearances and disappearances, and celestial positions. Later astronomers have found these old comet annals to be a great resource.

Comets, we now know, are remnants from the birth of our solar system, some 4.6 billion years ago, and are largely made up of ice coated in black organic stuff. "Dirty snowballs," as they've been dubbed.

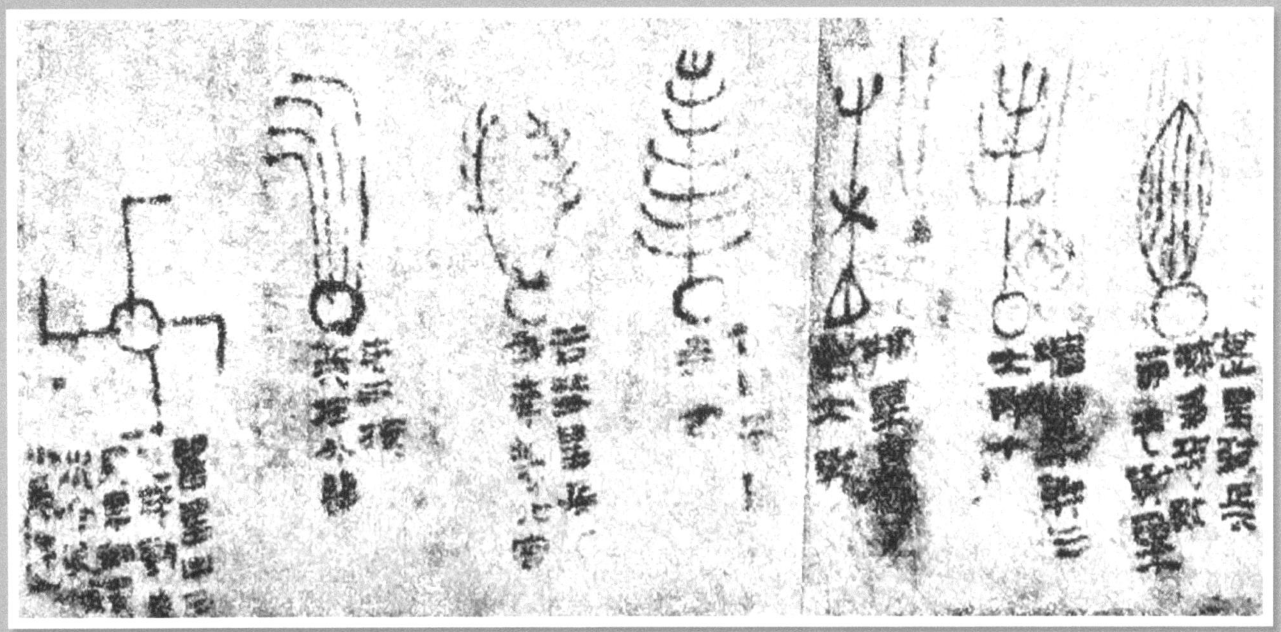

https://en.wikipedia.org/wiki/Historical_comet_observations_in_China

Comet 45P/Honda-Mrkos-Pajdušáková is captured using a telescope on December 22 from Farm Tivoli in Namibia, Africa.

Credits: Gerald Rhemann

https://www.nasa.gov/feature/goddard/2017/nasa-telescope-studies-quirky-comet-45p

They could provide crucial information regarding the genesis of our solar system. Water and organic molecules, the building elements of life, may have been transported by comets to the Earth and other parts of the Solar System.

What is the origin of COMETS?

As postulated by astronomer Gerard Kuiper in the year 1951, a disc-like belt of frozen bodies extends beyond Neptune, where a group of dark comets orbits the Sun in the Plutonian region.

✦ **These Short-Period Comets are ice objects which are occasionally forced into orbits that bring them closer to the Sun by gravity**. Because they orbit the Sun in less than 200 years, their emergence is often expected because they have gone through before.

✦ **Long-Period Comets**, which originate in a region called the Oort Cloud about 100,000 astronomical units (that is, 100,000 times the distance between Earth and the Sun) from the Sun, are less predictable. **These comets from the Oort Cloud could last up to 30 million years to complete one trip around the Sun.**

https://solarsystem.nasa.gov/resources/2527/comet-c2020-f3-neowise-over-utah/?category=small-bodies_comets Source: NASA/Bill Dunford

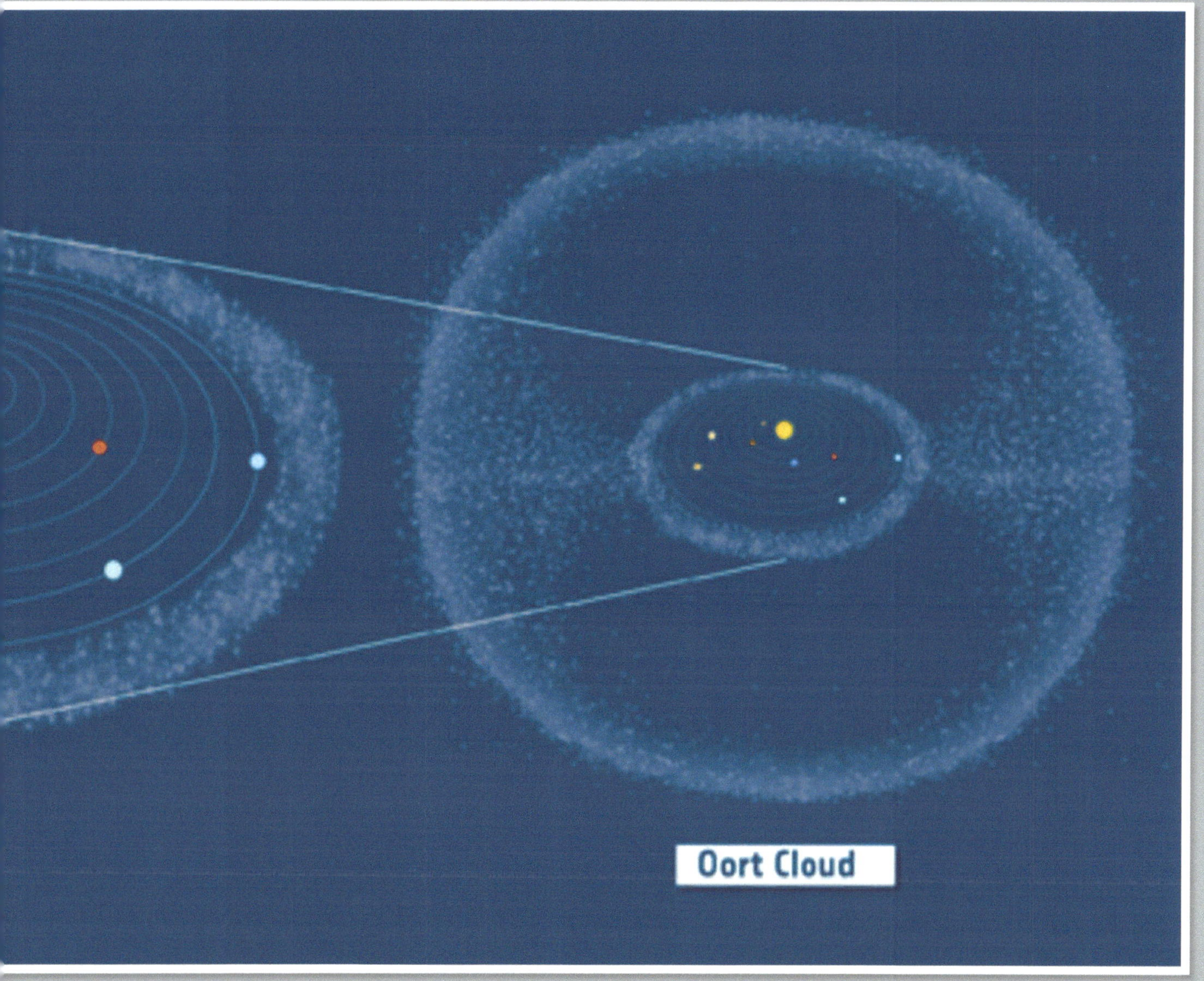

https://www.esa.int/ESA_Multimedia/Images/2014/12/Kuiper_Belt_and_Oort_Cloud_in_context

EACH COMET HAS A NUCLEUS, which is a little frozen portion that is often only a few kilometres in diameter. Icy chunks, frozen gases with dust imbedded in them can be found in the nucleus.

As a comet approaches the Sun, it warms up and forms a **COMA, OR ATMOSPHERE**. The heat from the Sun causes the ices on the comet to melt and turn into gases, expanding the coma. Thousands of kilometres may be covered by the coma.

Comets have TWO TAILS, one a DUST TAIL and another an ION (GAS) TAIL.

The coma dust and gas can be blown away from the Sun by the pressure of sunlight and high-speed solar particles (solar wind), generating a long, luminous tail in the process.

This infrared image from NASA's Spitzer Space Telescope shows the **broken Comet 73P/Schwassman-Wachmann 3** skimming along a trail of debris left during its multiple trips around the sun.

Most comets keep a safe distance from the Sun, with Halley coming no closer than 89 million kilometres (55 million miles).

Some comets, known as Sungrazers, crash into the Sun or get so close that they break up and evaporate.

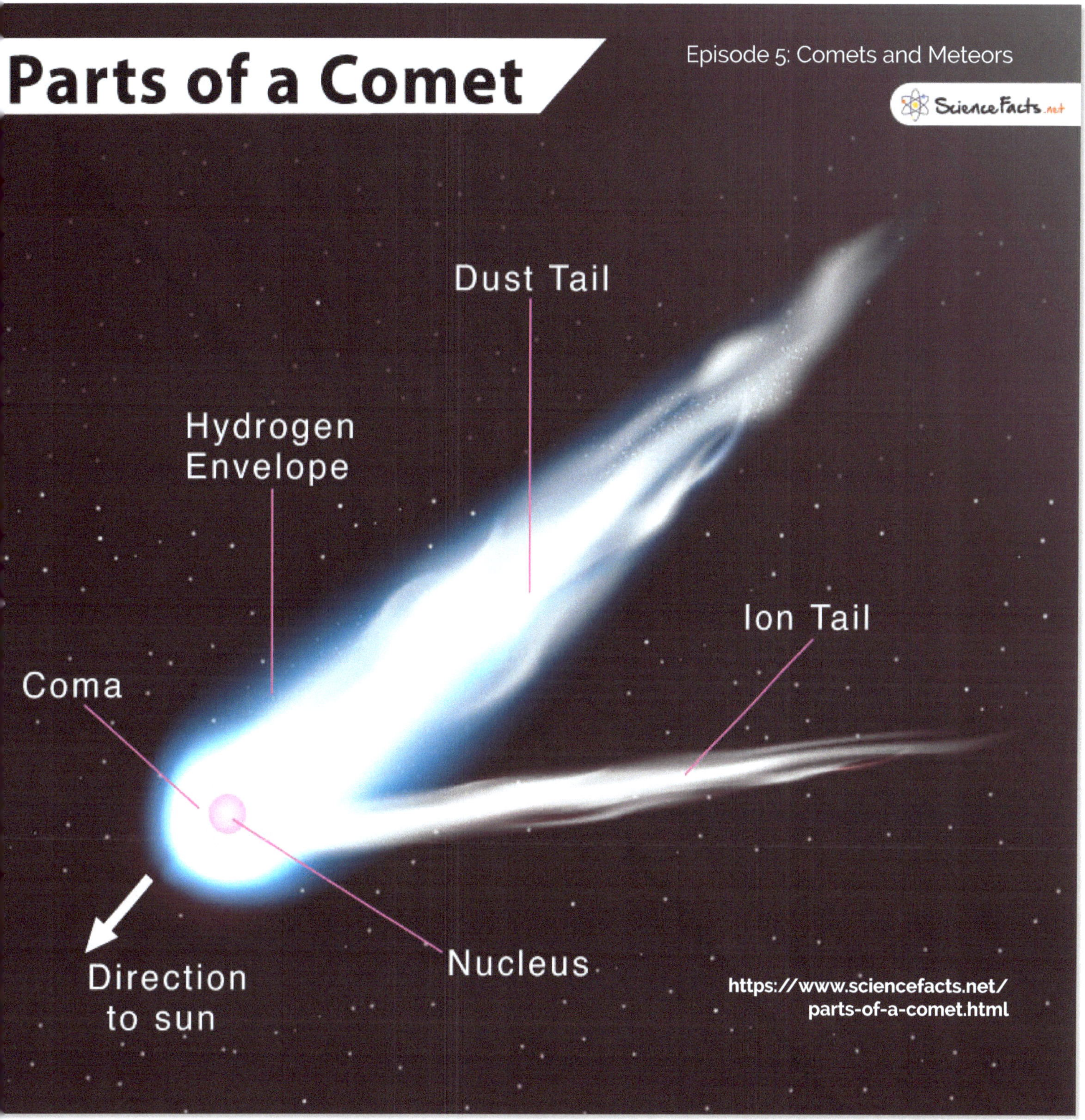

Parts of a Comet
Episode 5: Comets and Meteors
ScienceFacts.net
Dust Tail
Hydrogen Envelope
Ion Tail
Coma
Nucleus
Direction to sun
https://www.sciencefacts.net/
parts-of-a-comet.html

EXPLORATION OF COMETS

Scientists have long wanted to study comets in depth, enticed by the few images of comet Halley's nucleus taken in 1986. In 2001, NASA's Deep Space 1 spacecraft photographed the nucleus of comet Borrelly, which is about 8 kilometres (5 miles) long.

In January 2004, NASA's Stardust mission flew within 236 kilometres (147 miles) of the nucleus of Comet Wild 2, collecting cometary particles and interstellar dust for a sample return to Earth in 2006. The images captured during this close flyby of a comet nucleus show dust jets and a rugged, textured surface. The analysis of the Stardust samples indicates that comets are more complex than previously thought.

Minerals generated near the Sun or other stars were discovered in the samples, implying that components from the solar system's inner regions made their way to the outer regions, where comets formed.

Deep Impact was a NASA mission that included a flyby spacecraft and an impactor. The impactor was launched into the path of comet **Tempel 1's nucleus** in a scheduled collision in July 2005, vaporising it and ejecting large volumes of fine, powdery material from underneath the comet's surface. The impactor camera imaged the comet in greater detail as it

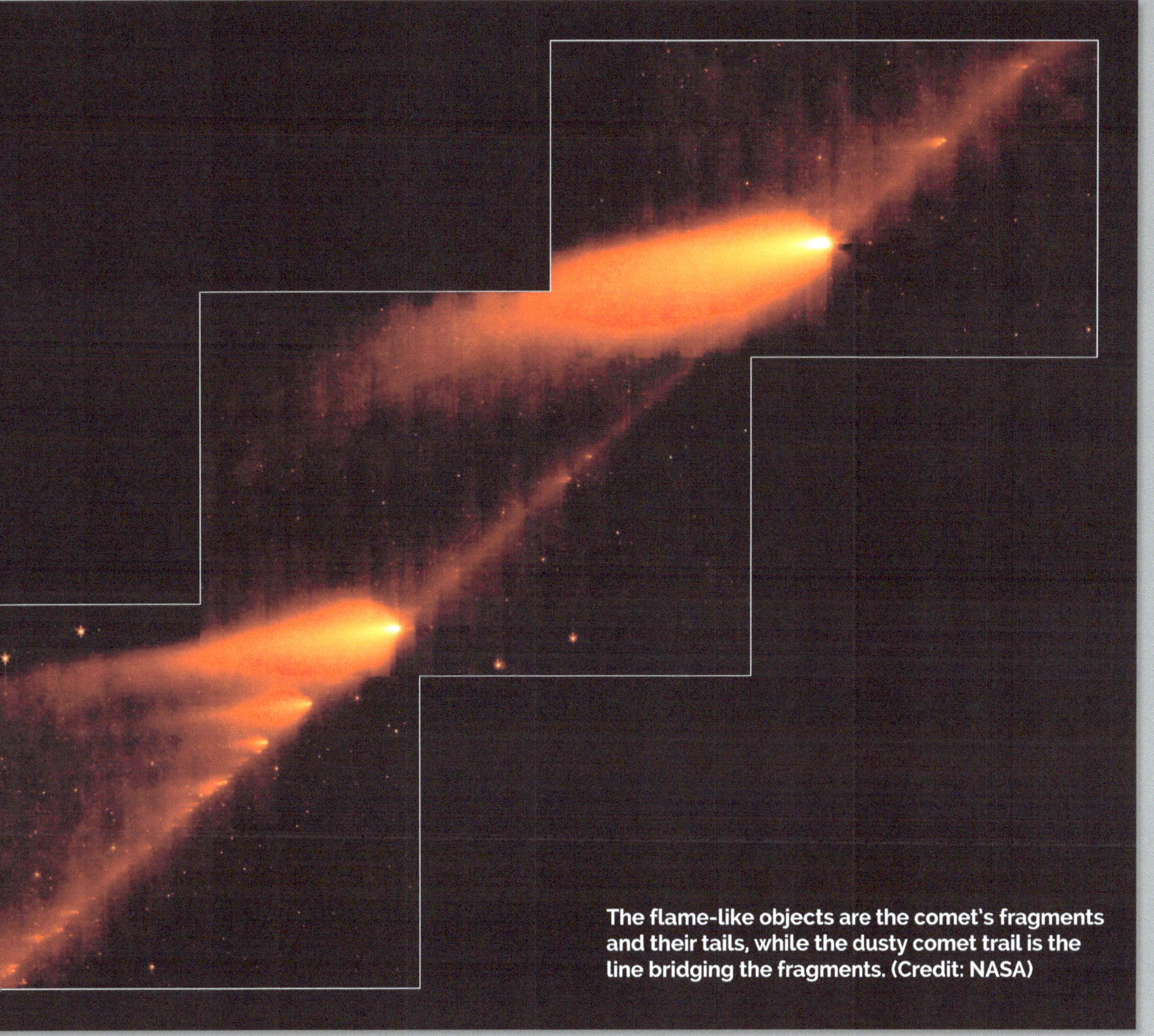

The flame-like objects are the comet's fragments and their tails, while the dusty comet trail is the line bridging the fragments. (Credit: NASA)

approached impact. The stunning excavation was captured by two cameras and a spectrometer on the flyby spacecraft, which helped identify the internal composition and structure of the nucleus.

The Deep Impact and Stardust spacecraft were still healthy after their successful original missions and were retargeted for additional cometary flybys.

EPOXI (Extrasolar Planet Observation and Deep Impact Extended Investigation) was Deep Impact's mission, and it consisted of two projects: the Deep Impact Extended Investigation (DIXI), which collided with **Comet Hartley 2** in November 2010, and the Extrasolar Planet Observation and Characterization (EPOCh) investigation, which looked for Earth-size planets around other stars on the way to Hartley 2.

The Stardust New Exploration of Tempel 1 (NExT) mission identified changes in the nucleus following Deep Impact's 2005 encounter in 2011, prompting NASA to return to Tempel 1.

How do Comets get Their Names?

The naming of comets can be difficult. **Comets are usually named after the person or spacecraft who discovered them.** This guideline from the International Astronomical Union was only developed in the recent century.

For example, **Comet Shoemaker-Levy 9** was named after Eugene and Carolyn Shoemaker and David Levy, who discovered it as the ninth short-periodic comet. Because spacecraft are so good at identifying comets, many comets have names like **LINEAR, SOHO, or WISE**.

Thus, if we consider those objects as comets, what are meteors?

As previously stated, a comet is composed primarily of ice, rock, gases, and dust and orbits the sun.

Meteors, on the other hand, are streaks of light seen when a space rock enters the Earth's atmosphere and begins to 'burn up.'

'STREAKS OF LIGHT' METEOROIDS, METEORS OR METEORITES?

As a result, the term "SHOOTING STAR" was coined. However, what you're actually seeing is a tiny piece of interplanetary debris: rock, ice, or metal slamming through the Earth's atmosphere and becoming incandescent. Most are dim, but some can be extremely bright.

But do we call these 'streaks of light' meteoroids, meteors or meteorites?

Let's find out.

✦ A **meteoroid** is a piece of solid material that falls from space.

✦ A **meteor** is the phenomenon of a meteoroid becoming hot and blazing across the sky.

✦ Finally, if it hits the ground, it is referred to as a **meteorite.**

Most meteors are about the size of a grain of dust. Even smaller sometimes. However, its enormous speed gives our meteor a vast kinetic energy as it pierces through the sky.

Meteors, which are hot and blazing through the sky, speed through our atmosphere at incredible speeds.

The meteoroid is likely orbiting the Sun at a few dozen kilometres per second.

As it approaches the Earth, the gravity of our planet accelerates it by 11 kilometres per second — the

Earth's escape velocity. When it enters our atmosphere, it travels at speeds of up to 70 km/sec or more.

Don't let its brightness and crazy speed fool you.

When they hit our atmosphere, they come to a halt from their ridiculous orbital speed, and all that energy has to go somewhere. It is converted into light and heat, which we see as a meteor.

Do you think that a meteor gets hot due to its friction with air?

Well, you're correct but there's a bigger boss to this. **Compression**. Now I'm sure we're aware of the fact that when we compress a gas, it gets heated up. Our meteoroid, that is coming at gigantic speeds, constantly compresses the gas in front of it a lot. So much that some of the air can reach temperatures of over a 1000 degrees celsius! The air soon radiates this heat which heats up our meteoroid. This enormous amount of heat vaporises some of the material and it blows away. And if you're still wondering how we get that 'streak of light' behind the meteor, you have your answer. This process of the material being blown away is termed as **ablation**.

(Notice how I said meteoroid and not meteor. Meteoroids revolve around the sun and not meteors.)

I think we can all agree that this 'shooting star' is pretty cool. But you know what's cooler? A METEOR SHOWER.

Every year, there are approximately 21 meteor showers, with the majority of them occurring between August and December.

**https://blogs.nasa.gov/Watch_
the_Skies/tag/meteor-shower/**

We can say that meteors are like wolves. Kinda. Sometimes they travel alone and sometimes they travel in packs.

When meteors travel in packs, you guessed it right, we get a meteor shower.

The number of meteors racing towards earth can sometimes exceed a hundred meteors per hour.

Some meteor showers don't come from asteroids. They come from comets.

Meteor showers of this type occur when the Earth's orbit intersects the orbit of a comet. Comets leave behind rocky trails that are often the size of pebbles or grains of sand but can be as large as boulders. Every year, the Earth passes through these debris' trails known as meteoroid streams, and the planet becomes strewn with rocky material.

Meteor showers are a reminder of our place in a dynamic and beautiful cosmic ecosystem, inspiring everything from making wishes to revelling in the sky.

In this 30 second exposure, a meteor streaks across the sky during the annual Perseid meteor shower Friday, Aug. 12, 2016 in Spruce Knob, West Virginia. The Perseids show up every year in August when Earth ventures through trails of debris left behind by an ancient comet. Image Credit: NASA/Bill Ingalls

EPISODE 6:
ASTEROIDS

What constitutes an asteroid and what does not constitute an asteroid is a bit of a grey area. **But, in general, it's a rocky or metallic class of smaller worlds that orbit the Sun out to Jupiter.**

The majority of this ancient space junk may be found in the main asteroid belt, orbiting our Sun between Mars and Jupiter.

Asteroids range in size from Vesta, the largest at 329 miles (530 kilometres) in diameter, to asteroids as small as 33 feet (10 metres).

The cumulative mass of all asteroids is less than that of the Earth's Moon.

Asteroids are available in a few different flavours. The majority of them, almost 3/4, are **carbonaceous**, meaning they contain a lot of carbon. About 1/6th is **silicaceous,** which means it's heavy in silicon-based components, like rock. **Metal objects**, literally filled with iron, nickel, and other metals, dominate the remainder, which are put into one catch-all category.

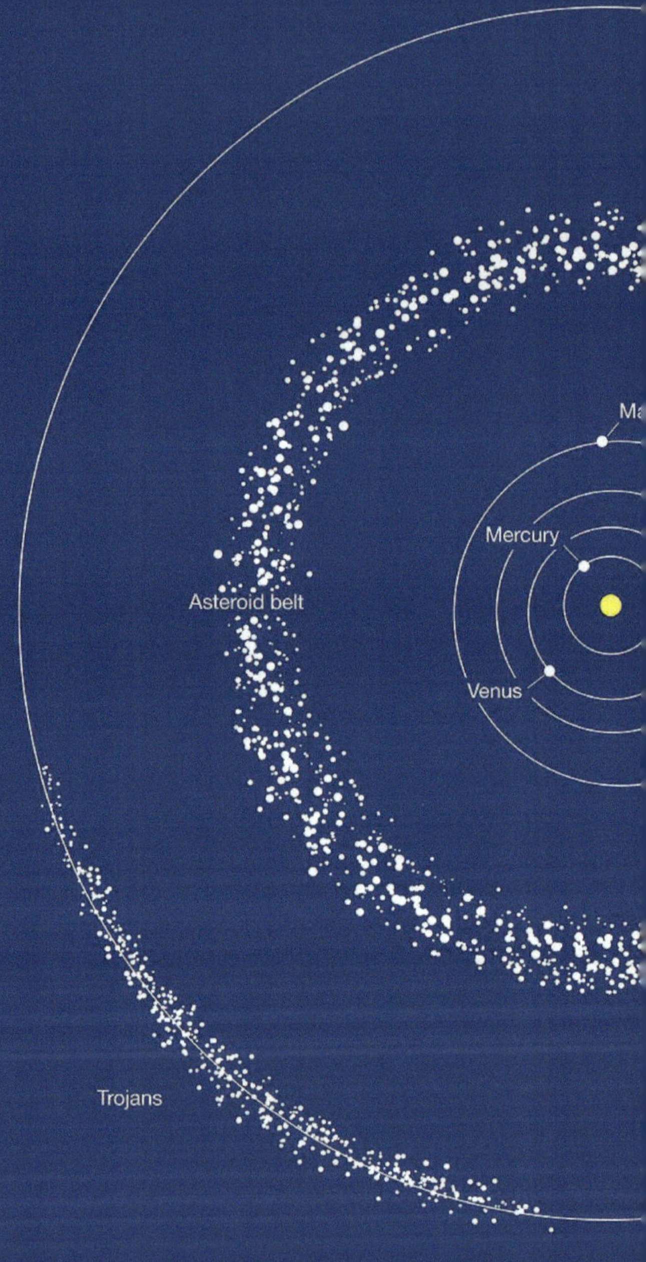

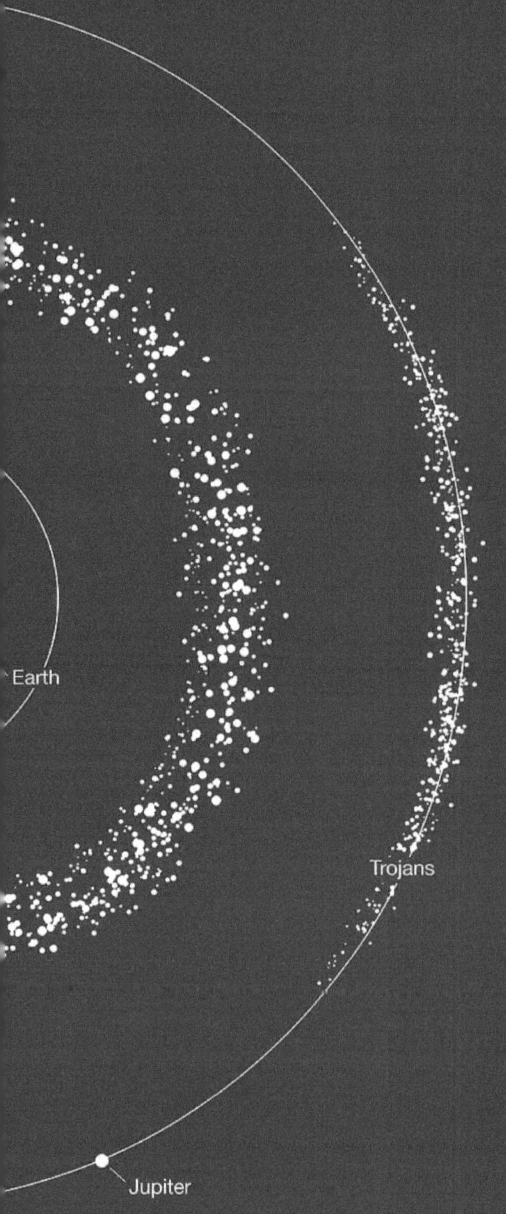

This image depicts the two areas where most of the asteroids in the Solar System are found: the asteroid belt between Mars and Jupiter, and the trojans, two groups of asteroids moving ahead of and following Jupiter in its orbit around the Sun.

The binary asteroid 288P is part of the asteroid belt.

https://esahubble.org/images/heic1715c/
Credit: ESA/Hubble, M. Kornmesser

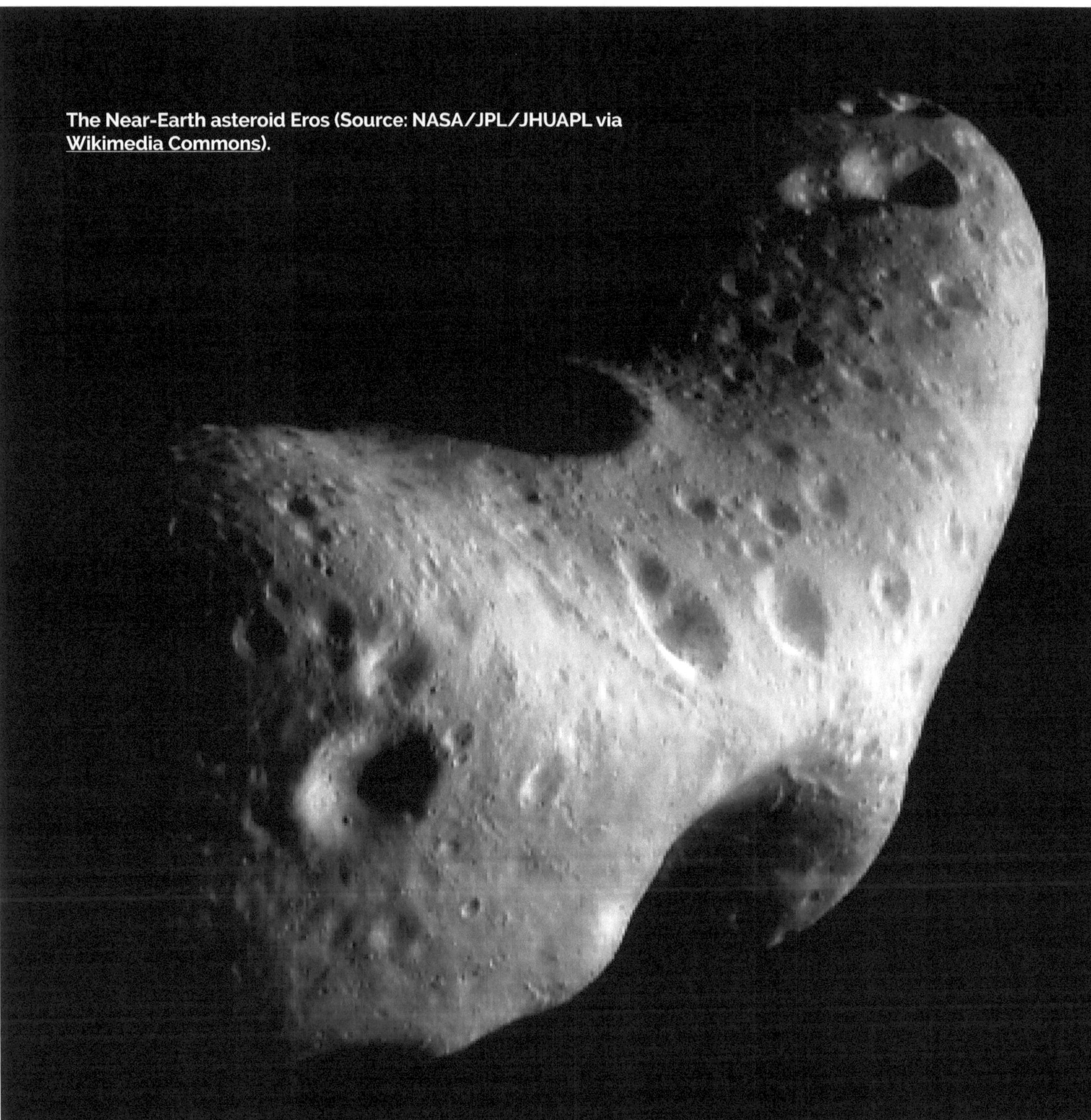

The Near-Earth asteroid Eros (Source: NASA/JPL/JHUAPL via
Wikimedia Commons).

Asteroids and comets have been crashing into Earth since the beginning of time. In fact, the most widely accepted idea for the extinction of dinosaurs is that they were wiped off by an asteroid that collided with Earth, wreaking massive havoc, and obliterating most life forms in the process.

Even while numerous astronomical objects approach our planet, many of them are burned up in the atmosphere and never reach the surface. Space rocks less than 25 metres would most likely burn up once they hit Earth's atmosphere, according to NASA.

Some larger asteroids, on the other hand, do make it through the atmosphere and strike the surface of our planet.

In 2013, an asteroid collided with the atmosphere over Chelyabinsk, Russia.

This event sent a shockwave through the city, injuring 1,200 people.

A global apocalypse would require an asteroid with a diameter of more than 400 metres. On a worldwide scale, an asteroid impact would bring tremendous destruction of both life and property.

Fortunately, such massive asteroids only strike Earth once per 1000 generations on average. Overall, asteroids can and do orbit close to our planet, but the vast majority will never approach, and even if they do, they will be burned up in the atmosphere before ever touching our planet's surface.

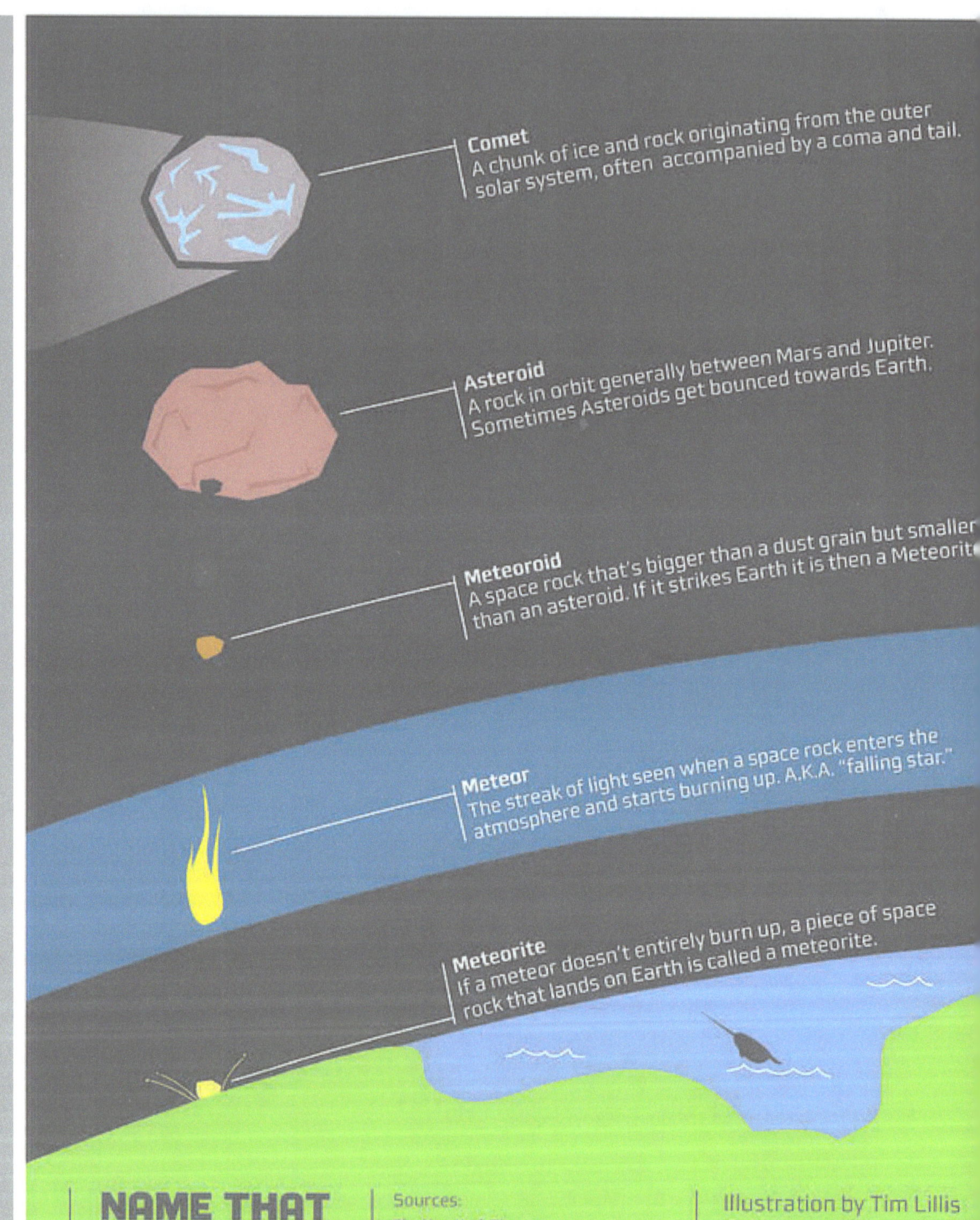
Comet
A chunk of ice and rock originating from the outer solar system, often accompanied by a coma and tail.
Asteroid
A rock in orbit generally between Mars and Jupiter. Sometimes Asteroids get bounced towards Earth.
Meteoroid
A space rock that's bigger than a dust grain but smaller than an asteroid. If it strikes Earth it is then a Meteorite.
Meteor
The streak of light seen when a space rock enters the atmosphere and starts burning up. A.K.A. "falling star."
Meteorite
If a meteor doesn't entirely burn up, a piece of space rock that lands on Earth is called a meteorite.
NAME THAT SPACE ROCK!
Sources:
The New York Times:
Was that Fireball a Meteor or a Meteorite?
Wikipedia pages for Asteroid and Comet
Illustration by Tim Lillis
@tim_narwhal
narwhalcreative.com

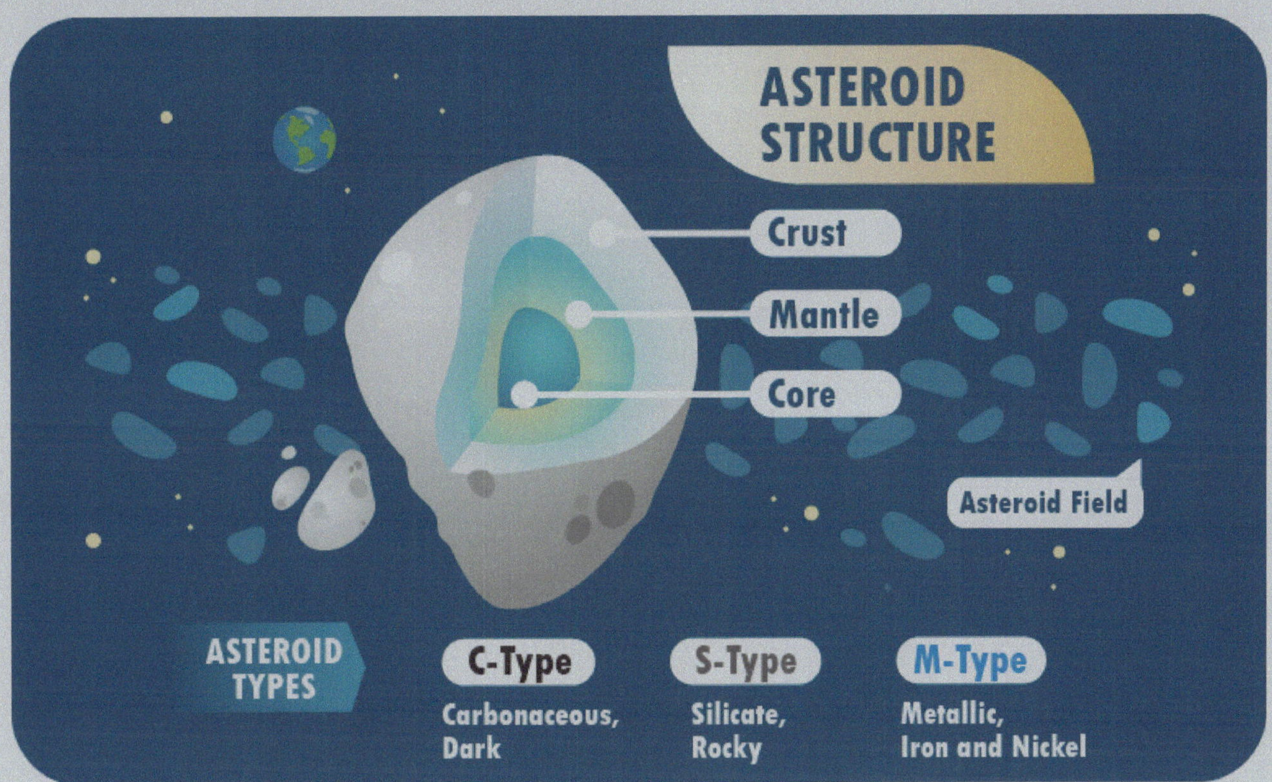

https://letstalkscience.ca/educational-resources/backgrounders/where-can-we-find-asteroids-and-comets

FUN FACTS OF
ASTEROIDS & COMETS

1. The total number of known asteroids is **1,113,527.**

2. Asteroids were originally named after **feminine goddesses, such as Ceres, Vesta, and Juno.** We ran out of names as hundreds, then thousands, more were discovered. After a lengthy proposal and acceptance process overseen by the International Astronomical Union, astronomers who found asteroids were eventually authorised to name them.

3. **Comet is derived from the Greek term Kometes, which means "long hair."** This is due to the fact that a comet's tail might resemble long strands of hair.

4. **Halley's Comet is the most famous comet of all time.** Halley is a periodic comet that has been observed from Earth for ages and is visible every 76 years. The last time it was seen was in 1986.

The Hale-Bopp Comet, Donati's Comet, and the Shoemaker-Levy 9 Comet are among the other well-known comets.

EPISODE 7: BLACK HOLES

The Final Destination: Behold, the final chapter.

We've spoken about a lot of fascinating stuff in this book. But for black holes, the word 'fascination' is an understatement. Maybe mind-blowing, eye-popping or outstanding could work.

Let's begin.

Ah, **BLACK HOLES.** The phrase is tossed around quite often. They're the subject of your favourite science fiction movies. Time travel to different universes, entrance portals through wormholes and what not. The way black holes have been portrayed; you must be thinking that black holes are some sort of a myth. Infinite density in zero volume? Nah, that can't be right.

Okay, but what really is a BLACK HOLE?

Is it some sort of a star?

Is it a certain kind of galaxy?

Is it a portal to a universe where you are a billionaire who owns 73 different types of cars?

Well, I'm sorry to break it to you, but black holes are comparatively simpler to define.

There is no way back once something crosses the **'EVENT HORIZON,'** as it is known. We can't see black holes because they are so powerful that even light is dragged in.

A black hole is a region of spacetime where gravity is so powerful that nothing — no particles or even electromagnetic radiation such as light — can escape from it.

The term "black hole" is self-explanatory enough.

Matter has collapsed in itself, and you feel as if you're staring into the void. The term "black hole" is simple enough but picturing one in space is difficult. Consider a massive drain with water cascading down into it.

NOW LET'S SEE HOW THESE MASSIVE GIANTS ARE FORMED.

Flashback! Do you remember what happens after stars die? Don't worry, I'm not going to quiz you. Here's a tiny recap.

Strange remnants remain after a star dies: stars like our Sun make white dwarf stars, whereas

Credits: NASA/CXC/M.Weiss

An artist's drawing a black hole named Cygnus X-1. It formed when a large star caved in. This black hole pulls matter from blue star beside it.

other high mass stars produce neutron stars. Instead of decaying slowly, high-mass stars burn through their fuel quickly and explode in enormous supernova explosions. **If the star is huge enough, its death results in an even stranger thing. A BLACK HOLE.**

Not all black holes are created when stars explode at the end of their lives.

Only stars significantly more massive than our Sun can generate black holes.

However, there are still many open questions about how the black holes form. For instance, how massive must a star be in order for it to form a black hole? The standard limit that is usually quoted is that the mass when it was a main sequence star should be larger than 30 solar masses. However, this limit is not really known that accurately. It could be a bit smaller or larger.

A BLACK HOLE is created when the mass of an object collapses into an infinitely small point.

In their final stages, enormous stars go out with a bang in massive explosions known as SUPERNOVAE. Such a burst flings star matter out into space but leaves behind the stellar core. While the star was alive, nuclear

https://www.forbes.com/sites/jamiecartereurope/2022/03/28/this-jaw-dropping-supernova-explosion-nebula-may-have-collided-with-something-say-scientists/?sh=23fec8e03272

BLACK HOLES' DIET MAINLY COMPRISES GAS AND DUST

fusion created a constant outward push that balanced the inward pull of gravity from the star's own mass. In the stellar remnants of a supernova, however, there are no longer forces to oppose that gravity, so the star core begins to collapse in on itself.

Black holes' tremendous gravitational pull comes from packing all of that mass—many times the mass of our own sun—into such a small space. There could be tens of thousands of these stellar-mass black holes in our own Milky Way galaxy.

Okay, but all black holes aren't the same. #NotAllBlackHoles (Sorry). Some are babies. And some are giants.

Infrared Echoes of a Black Hole Eating a Star

https://www.nasa.gov/image-feature/jpl/ pia20027/infrared-echoes-of-a-black-hole-eating- a-star Image credit: NASA/JPL-Caltech

These baby-black holes also known as **Stellar Mass Black Holes** are formed by the collapse of huge stars and are roughly 10-24 times the mass of the Sun. When a star approaches this baby black hole close enough for some matter to be snared on by the black hole, the **Accretion Disc** of the black hole produces, and get this, **X-RAYS.**

This is how we detect these babies. Scientists predict that there are as many as ten million to a billion such black holes in the Milky Way alone, based on the number of stars massive enough to form them.

Now let's come to the giants. These giants, **A.K.A. Supermassive Black Holes** have a whopping mass of millions of solar masses. They exist at the heart of every major galaxy, even our own milky way. Astronomers successfully detect these giants by the effects they have on neighbouring gas and stars due to their immense gravitational pull.

You may be wondering, what about teenagers? I mean, what about middle-sized black holes? Astronomers have long assumed that there are no mid-sized black holes. On the other hand, recent data from Chandra, XMM-Newton, and Hubble reinforces the case for the presence of mid-sized black holes. According to one scenario, a chain reaction of star collisions in dense star clusters produces extremely massive stars which eventually collapse to produce intermediate-mass black holes. The star clusters subsequently make their way to the galaxy's core, where the intermediate-mass black holes unite to become a supermassive black hole.

Quick question, what did you have for breakfast today? Well, whatever it was, I'm sure it wasn't what our black holes feast upon.

Anything that comes too close to a black hole is sucked in so strongly that it has no possibility of escaping. (Maybe like a McDonald's burger just before you take a bite from it).

While black holes do have event boundaries, and anything that crosses them cannot escape, **they eat more like a two-year-old, the messiest eaters.**

This material being drawn in is what causes a black hole to expand in size.

Astronomers have discovered pools of cool gas around some of the Universe's earliest galaxies. These are the ideal meal for the supermassive black holes at the heart of these galaxies to aid their growth.

Gas and dust are fine but what really satisfies the appetite of our hungry black holes? You guessed it right. Stars.

Black holes can have low mass or high mass companion stars. How mass is exchanged, and hence how the black hole is fed, is determined by the sort of companion star. Large amounts of mass can be transferred into the black hole by high-mass stars' stellar winds. Low-mass stars, on the other hand, have significantly weaker winds, necessitating dangerously close proximity in order for them to feed the black hole.

Consider the fact that a black hole prefers to eat either stars or gas clouds.

A typical gas cloud in space is many times the size of our Solar System, with some reaching many light years, but a star approaching a black hole will get spaghettified, or stretched into a long, thin thread aligned with the black hole's direction.

By the time either of these possibilities reaches the black hole's event horizon, they've grown to be several times its size.

Instead, just a small bit of the cookie reaches the event horizon, much like when a two-year-old eats a cookie.

Due to tremendous gravitational pressures and the large size mismatch between the comparatively tiny black holes and the massive clumps of matter that feed them, the vast majority of infalling matter is spat back out in an explosive, violent flurry.

Contrary to popular assumption, upwards of 90% of infalling matter will never make it into a black hole, according to estimations. Instead, it's thrown out into the outer reaches of the galaxy, where it can drive the development of new stars before finally returning to the interstellar medium.

Black holes are among the universe's worst eaters. So, don't be discouraged if you've ever witness a kid eat a fourth of her food and then dumps the rest all over her face, the table, and the floor. She is a lot better than a black hole at the very least.

This food storage may explain why black holes evolved so swiftly early in the history of our Universe.

Now, let's learn about a process called ACCRETION.

WHAT IS ACCRETION?

You know how during lunch time in school your friend group comes together? Similarly, **Accretion is the term used in astrophysics to describe the process of gas and dust being gathered together by gravitational forces**.

Accretion can take place on a variety of scales. For example, in the early history of our own solar system, gravity drove the accretion of smaller particles into larger and larger ones, eventually developing from tiny grains of dust to planetesimals and then to the planets we know today.

So, what is its Significance?

Accretion is also responsible for the formation of stars from hydrogen clouds, as well as the formation of galactic discs from stars.

https://science.howstuffworks.com/accretion-disk.htm

https://www.universetoday.com/142453/black-hole-simulation-solves-a-mystery-about-their-accretion-disks/

Accretion is also the process through which black holes are nourished. When a star or a cloud of material passes close enough to a black hole, it is subjected to the black hole's gravitational effects, and the force of attraction drives material towards the black hole

And now, behold

Welcome to the black hole! Now that we have arrived in orbit near a black hole's event horizon, it's time to dip down into the interior.

The event horizon a strange one-way street in the universe, preventing light and material that have crossed from returning.

The appearance of an event horizon—a border through which matter and light may only pass inward towards the black hole's mass—is a defining feature of a black hole.

Inside the event horizon, nothing, not even light, can escape. Because information from an event that occurs within the event horizon cannot reach an outside observer, it is difficult to tell whether such an event occurred.

Although black holes do not have a "surface", scientists have defined a boundary to separate the "interior" of the black hole from its "exterior". A black hole's event horizon is a boundary that separates the black hole's interior, which we are unable to see, from the outer region. **Therefore, we say that the event horizon is the "surface", or boundary of a black hole, not as a rigid body, but as the point-of-no-return for material that has fallen in.**

Now let's do something unwise and travel past the Event Horizon of a black hole.

Once we pass through the event horizon we will be in the strange world of a black hole's interior. Although it is impossible to send information about the inside of the black hole to the universe beyond the event horizon, there are no laws of physics that would prevent us, observers within the event horizon, from making scientific discoveries.

The first thing that we would notice looking away from the black hole, is all of the light emitted by the stars and galaxies outside of the black hole.

It definitely isn't black inside a black hole!

If we shine a flashlight, we find that no matter what direction we try to aim the light, the rays always end up pointing inward, to smaller values of the black hole's radius.

A gravitational **SINGULARITY,** a region where the spacetime curvature becomes infinite, may exist in the heart of a black hole, as described by general relativity.

This region has the shape of a single point for a non-spinning black hole, and it is smeared out to create a ring singularity in the plane of rotation for a rotating black hole. The singular region has zero volume in both circumstances.

It's also possible to establish that the singular region includes all of the black hole solution's mass. As a result, the singular region can be considered to have infinite density.

Observers falling into a **Schwarzschild Black Hole** (i.e., one that is non-rotating and uncharged) will be carried into the singularity once they cross the event horizon.

They are crushed to infinite density when they approach the singularity, and their mass is added to the black hole's mass. Before that happens, they will have been torn apart by a process sometimes referred to as **Spaghettification or the "Noodle Effect".**

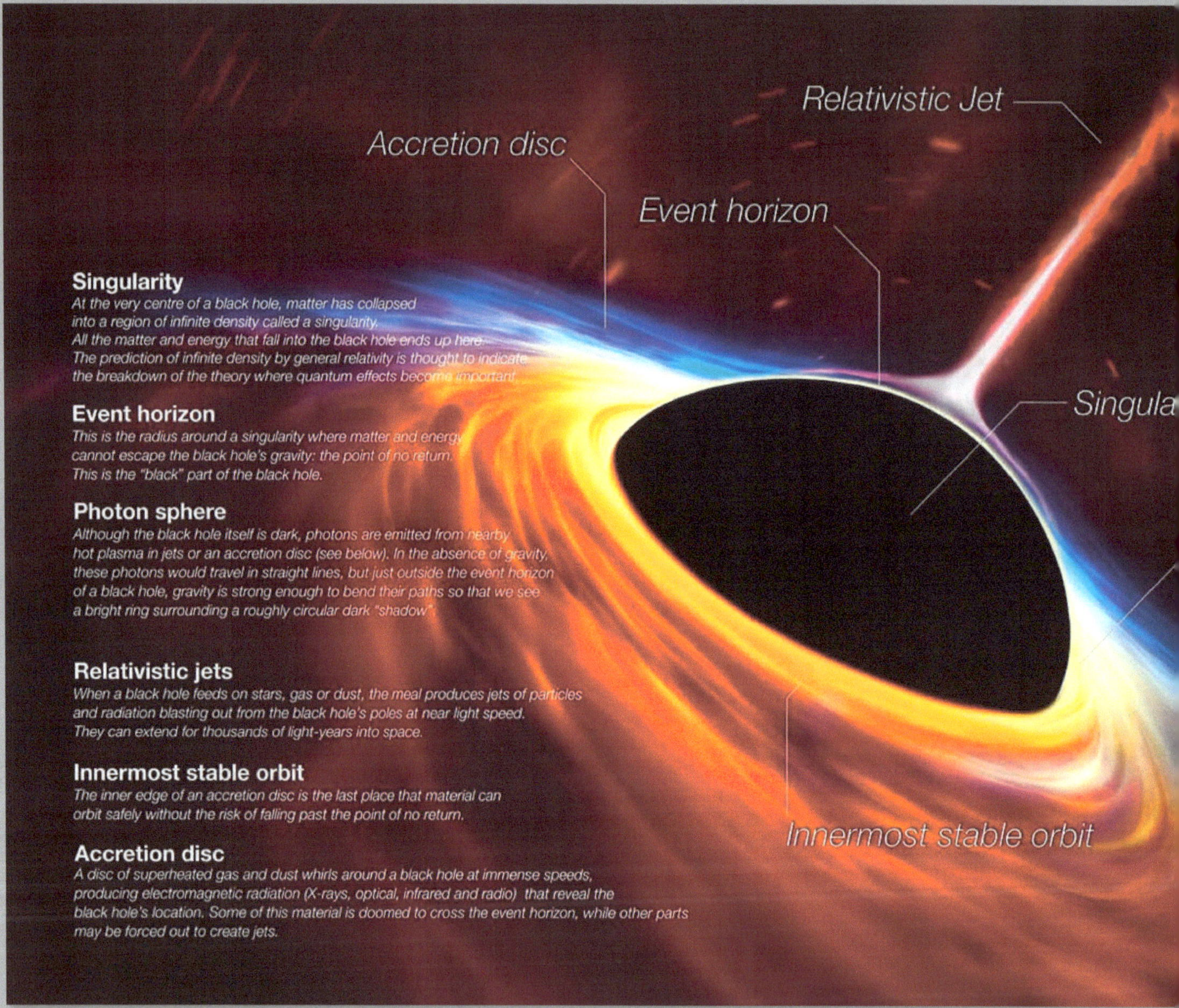
Relativistic Jet
Accretion disc
Event horizon
Singula
Innermost stable orbit
Singularity
At the very centre of a black hole, matter has collapsed into a region of infinite density called a singularity. All the matter and energy that fall into the black hole ends up here. The prediction of infinite density by general relativity is thought to indicate the breakdown of the theory where quantum effects become important.
Event horizon
This is the radius around a singularity where matter and energy cannot escape the black hole's gravity: the point of no return. This is the "black" part of the black hole.
Photon sphere
Although the black hole itself is dark, photons are emitted from nearby hot plasma in jets or an accretion disc (see below). In the absence of gravity, these photons would travel in straight lines, but just outside the event horizon of a black hole, gravity is strong enough to bend their paths so that we see a bright ring surrounding a roughly circular dark "shadow".
Relativistic jets
When a black hole feeds on stars, gas or dust, the meal produces jets of particles and radiation blasting out from the black hole's poles at near light speed. They can extend for thousands of light-years into space.
Innermost stable orbit
The inner edge of an accretion disc is the last place that material can orbit safely without the risk of falling past the point of no return.
Accretion disc
A disc of superheated gas and dust whirls around a black hole at immense speeds, producing electromagnetic radiation (X-rays, optical, infrared and radio) that reveal the black hole's location. Some of this material is doomed to cross the event horizon, while other parts may be forced out to create jets.

https://www.snmstore.com/?category_id=5075334

Interesting Q&A About BLACK HOLES

If nothing can escape from a black hole, then won't the whole universe eventually be swallowed up?

The universe is enormous. In comparison to the size of a galaxy, the size of a region where a particular black hole exerts considerable gravitational influence is quite small.

This is true even for supermassive black holes like the one found in the Milky Way's centre — **Sagitarrius A.** Most or all of the surrounding stars have been "eaten" by this black hole, whereas stars further out are relatively safe from being drawn in.

Because this black hole currently weighs a few million times the mass of the Sun, swallowing a few additional Sun-like stars will only result in minor mass increases.

The Earth 'which is 26,000 light years away from the Milky Way's black hole' is not in any danger of being sucked in.

Black holes will grow in size as a result of future galaxy collisions, such as the merging of two black holes. However, because the universe is large and expanding, collisions will not continue endlessly, and any form of black hole runaway effect is quite unlikely.

BLACK HOLES DON'T SUCK

Some people believe that black holes are **cosmic vacuums** that suck the space around them, but they are actually much like any other object in space, albeit with an extremely strong gravitational field.

Earth would not be sucked in if the Sun were replaced by a black hole of similar mass; it would continue to orbit the black hole as it does the sun now.

Although black holes appear to be pulling in matter from all directions, this is a widespread misconception. Companion stars give up some of their mass in the form of stellar wind, which eventually falls into the grasp of its hungry neighbour, a black hole.

SUPERMASSIVE BLACK HOLES ARE SAFER? WHAT?

If you're approaching a black hole, you're most probably falling. The only thing you need to figure out is whether all of you is falling at the same rate. Black holes are matter concentrated into an infinitely small point.

Your body begins to fall at varied rates as you approach the black hole feet first. The black hole tugs stronger at your feet than it does at your head, much as the moon tugs harder at the side of the Earth closest to it, causing the tide to rise up towards it.

The difference in pull is similar to being attached to two different cars, both travelling in the same direction but at different speeds: one at seventy miles per hour and the other at thirty miles per hour.

When your body is extended into a long strand of mush, it is affectionately referred to as "spaghettification."

Approaching the singularity at the heart of a black hole will inevitably result in spaghettification, but some black holes are more accessible than others. A black hole's event horizon is the point beyond which nothing returns.

Let's pretend you stepped directly on the horizon. It's not how fast you're falling that matters but how rapidly your head is falling in comparison to your feet.

With an event horizon the size of your body, a little black hole will exert significantly more force on your feet than on your head. A black hole ten thousand times your size will exert enormous force on both your head and your feet, but the difference in force will be insignificant.

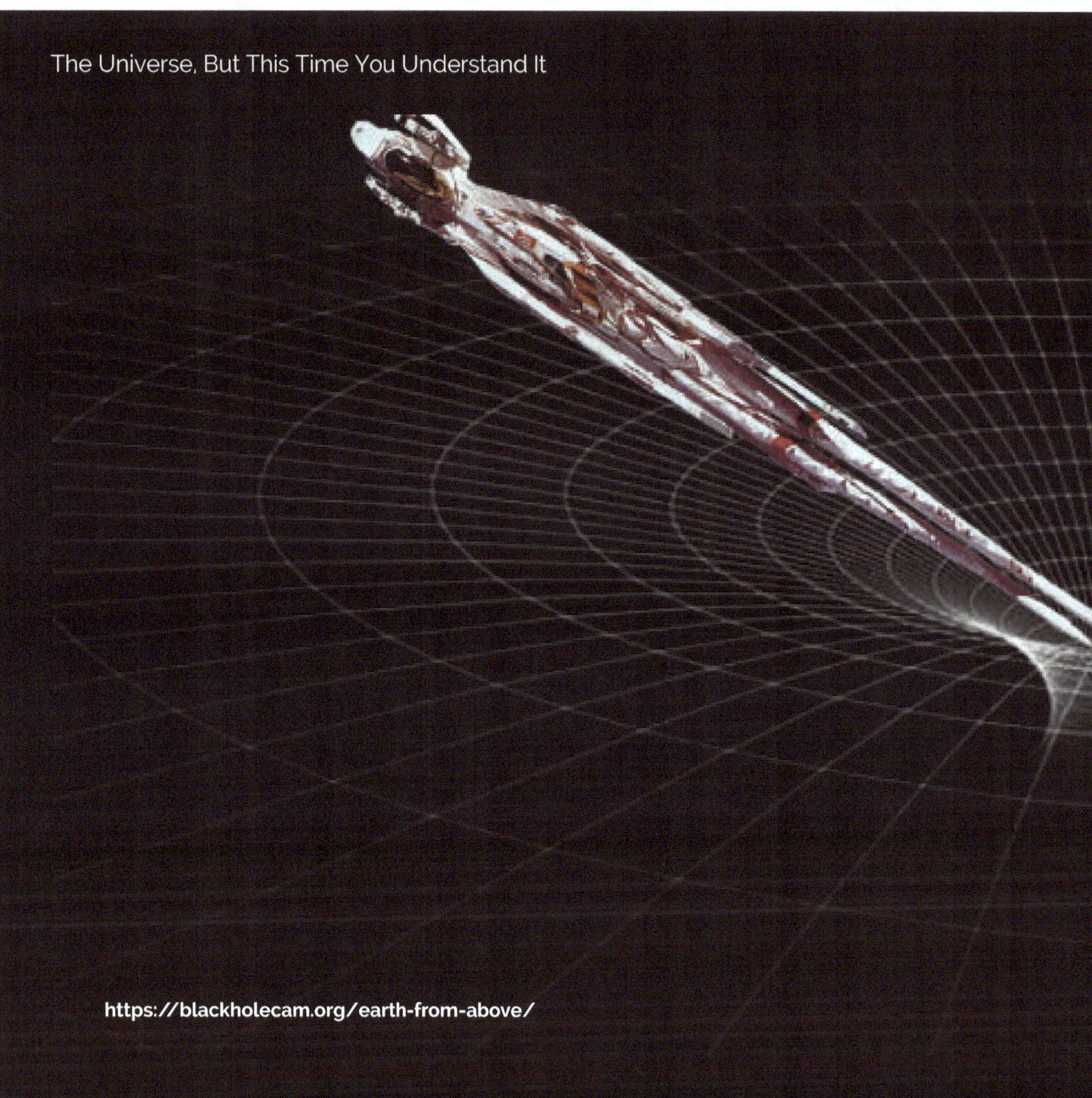

https://blackholecam.org/earth-from-above/

While both supermassive and tiny black holes will eventually rip you apart, the moment at which a little black hole will spaghettify you is well beyond its event horizon, and the point at which a big black hole will do damage is well within it. If you want to come near to a black hole but don't want to be torn apart like tissue, this is the way to go.

Anything can become a black hole.

The sole difference between a black hole and our Sun is that a black hole's center is formed of highly dense material, giving it a powerful gravitational pull. It is because of this gravitational field's ability to cage everything, including light, that we are unable to view black holes.

Anything can hypothetically become a black hole.

If our Sun were shrunk to a size of only 3.7 miles (6 kilometres) wide, for example, all of its material would be compacted into an incredibly small region, making it extremely dense and forming a black hole. The same notion might be applied to the Earth or your own body.

However, we only know of one means to create a black hole in reality: the gravitational collapse of a big star 20 to 30 times more gigantic than our Sun.

That was a lot, right? We went from the Big Bang to stars, galaxies, planets, meteors, comets, asteroids, and even black holes.

Well congratulations! You made it this far. This marks the end of our journey into the adventure of the cosmos.

We've always been explorers and we will always be!

ABOUT THE AUTHOR

Aashni Joshi, an aspiring Astrophysicist, and passionate Guitarist, brings to you her first book — The Universe, But This Time You Understand It.

From Teaching Physics to contributing on a NASA funded project — SUNGRAZER to Researching as an Intern at the Indian Institute of Technology — Bombay, Aashni is more than passionate about Physics. If she's not doing something Astrophysics related, she's probably watching a stand-up comedy show.

Her book not only brings out her love for Astrophysics but also how she incorporates humour into her work.